CONSENSUS ON PRE-COMMISSIONING STAGES FOR COGENERATION AND COMBINED CYCLE POWER PLANTS

AN ASME RESEARCH REPORT

Prepared by the
Combined Cycle Task Group
For the Water Technology Subcommittee of
the ASME Research and Technology
Committee On Water and Steam in Thermal
Systems

⊓ <u>FOREWORD</u> ⊓

The Water Technology Subcommittee of the ASME Research and Technology Committee on Water and Steam in Thermal Systems has established a Consensus on Pre-commissioning Stages of Cogeneration and Combined Cycle Power plants.

This publication is an important adjunct to several previously published documents prepared to inform, educate and assist the reader in adequately considering and planning for the many major activities involved in the design, construction and start-up of cogeneration and combined cycle power plants. Experience has shown that failure to consider the complex interrelationships between the various component systems can result in costly delays in project completion and turn over dates. Additionally, incorrect procedures and planning can result in damage or failure of key pieces of equipment or systems during subsequent operation.

This consensus was prepared by a task group of this subcommittee under the leadership of Edward Beardwood. The task group consisted of representatives of manufacturers, operators and consultants involved with the planning, design, manufacture, operation and monitoring of industrial and utility boilers, steam generators and associated equipment and systems.

The ASME Research and Technology Committee on Water and Steam in Thermal Systems will review, revise and reissue this document from time to time as necessary to comply with advances in technology in the design of these plants and the water treatment options.

Roger W. Light
Chair, ASME Research and Technology
Committee on Water and Steam in
Thermal Systems

❒ <u>ACKNOWLEDGEMENTS</u> ❒

This document was prepared by the Pre-commissioning Task Group of the Research and Technology Committee on Water and Steam in Thermal Systems of the American Society of Mechanical Engineers. Recognition is given to the following members of these groups for their contributions in the preparation of this document.

Edward Beardwood, Chair

Contributors:

A. Banweg

R. Bartholomew

E. Beardwood

W. Bernahl

J. Bellows

D. M. Bloom

I. J. Cotton

D. Daniels

D. Dewitt-Dick

J. C. Dromgoole

F. Gabrielli

R. T. Holloway

J. Isaac

C. M. Layman

R. W. Light

P. Midgley

W. Moore

J. Robinson

M. Rootham

J. Schroeder

K. A. Selby

D. E. Simon II

K. Sinha

M. Willett

This document is dedicated to the late Michael Rootham, a valued member of the ASME Water Technology Subcommittee and contributor to this consensus. His friendship, humor, wit, intellect, and courage will never be forgotten.

This document is intended to serve as a guide for planning for engineering, operations, maintenance, construction, and commissioning personnel designing and commissioning cogeneration and combined cycle power plants. It has been developed by the committee based on best practices and lessons learned from numerous experiences with the design, construction, commissioning, and operation of these types of plants. This document, however, is not a replacement for and does not supersede the procedures and recommendations of the original equipment manufacturers (OEMs) or the commissioning contractor. All procedures and requirements of the OEMs and the commissioning contractor must be adhered to in order to ensure validity of system equipment warranties.

❐ CONTENTS ❐

◗ PREFACE ◖

The ASME Research and Technology Committee on Water and Steam in Thermal Systems previously published a "Consensus on Operating Practices for the Control of Feedwater and Boiler Water Chemistry in Modern Industrial Boilers", ASME Publication CRTD – 34. That document has been widely applied worldwide by designers and operators of boilers and steam generators. It is used to establish critical monitoring and control parameters and normal operating limits that will minimize deposits and corrosion in the boiler and subsequent steam driven equipment.

The use and adoption of that document encouraged the Research Committee to undertake publication of the following documents that have proven to be of great assistance to industry.

- Consensus on Operating Practices for Control of Water and Steam Chemistry in Combined Cycle and Cogeneration Plants – ASME 859988
- A Practical guide to Avoiding Steam Purity Problems in the Industrial Plant – CRTD Vol. 35
- Consensus on Operating Practices for the Sampling and Monitoring of Feedwater and Boiler Water Chemistry in Modern Industrial Boilers – CRTD Vol. 81
- Consensus for the Lay-up of Boilers, Turbines, Turbine Condensers and Auxiliary Equipment – CRTD Vol. 66

The ASME Research and Technology Committee on Water and Steam in Thermal Systems has prepared this document to assist the industry in improving pre-commissioning practices for combined cycle power plants. This document is a consensus developed by the Pre-Commissioning Task Group, with input from manufacturers, engineering companies, and owners of combined cycle power plants.

Based on the committee members' experiences with delays in completion, start-up, and operation of new plants, it was concluded that publication of a document on pre-commissioning would be of interest and value. The committee is pleased to provide this compilation of best practice recommendations based on member's observations and experience.

This consensus is the minimum recommended practice. All requirements found within this document are to be fully monitored and recorded.

□ Section 1 □

INTRODUCTION

The time required to commission a combined cycle power plant can vary substantially and affect the overall cost of the project prior to commercialization. A number of pre-commissioning delays have been experienced in the independent power producing industry which are associated with the critical path of these projects. In some cases, important issues are overlooked or compromised in order to move to the next segment of the pre-commissioning. Shortcuts taken at various stages of the project, such as not electing to chemically clean the unit, can threaten the integrity of the equipment and/or delay the completion of the project. One survey of combined cycle pre-commissioning experiences revealed that 4 out of 22 performance issues accounted for 77% of the concerns. These were:

- Improper use of, or ignoring, standard practices and procedures, 27%
- Insufficient staff to accomplish the work in the scheduled time frame, 22%
- Poor project planning, 16%
- Breakdown in communication between project team members, 12%

It is critical that the availability of the water to be used for plant processes and its conditioning be well understood in the design stage of the project. The duration of the project commissioning process and the long-term reliability of the equipment can be affected by the ability to supply or produce water of the required quality or purity. Therefore, the production of high-purity water and the associated treatment issues must be considered part of the critical path. The chemistry criteria required to establish and maintain internal surface cleanliness during all stages of the project must not be overlooked.

The engineering, procurement, and construction (EPC) contractor's technical team should be involved in all phases of design specification development prior to the release of the RFQs (Request for Quotations). A list of preferred OEMs, fabricators/shops, and contractors should be developed based upon past experience, performance, and references. It should not be assumed that all bidders are equal, especially in terms of quality, worksite and job cleanliness. The EPC technical team should audit bidders to establish acceptability and be members of the bid (RFQ) review team.

EPC contractors can employ the techniques specified herein to ensure the chemistry requirements and limits are met during commissioning and through initial start-up operations. Operating for significant periods of time outside the chemistry limits imposed by heat recovery steam generator (HRSG) and turbine manufacturers will void the manufacturer's warranty. This will leave the Owner/ EPC contractor liable for the full cost of any repair that may be required within the warranty period.

During the plant design phase it may be beneficial and cost effective to consider additional equipment or modifications to equipment designs, which could significantly reduce the cost and duration of the start-up process even at the expense of increased capital cost. Post start-up operating problems and plant component failures may also be avoided in the long term. Examples of design phase improvements would be the inclusion of condensate polishers for start-ups, condenser leaks and equipment design upgrades for cycling operations.

❐ Section 2 ❐
SCOPE

This document provides guidance on design, procurement, and pre-commissioning activities that will result in the construction of a plant with steam/water-wetted surfaces that are as clean and corrosion-free as practical.

Issues can surface during the commissioning of a combined cycle power plant that cause unintended delays, cost overruns, increased post start-up maintenance, and depreciation of equipment. Consensus recommendations have been developed to minimize these risks and improve long-term reliability[1,2,3,4,5,6,7,8].

These procedures will be useful for those involved in the design, construction, and commissioning of a combined cycle power plant. This consensus is also applicable to cogeneration plants that export steam for use in a host process. All references to pH and conductivity assume they are measured at 77°F (25°C).

A generalized representation of a combined cycle plant development, construction and commissioning timeline is given in Figure 1. While this document primarily addresses the physical pre-commissioning stages, design specifications are also very important, and equipment selection determines the available cycle chemistry options. These concerns are discussed in the document and referenced to related literature sources. Influent water quality, treatment and the resultant water purity are important parts of design and pre-commissioning. Poor equipment cleanliness and improper water chemistry have been responsible for costly delays and corrosion-related failures[9,10].

Figure 1: Combined Cycle Plant Development, Construction and Commissioning Time Line

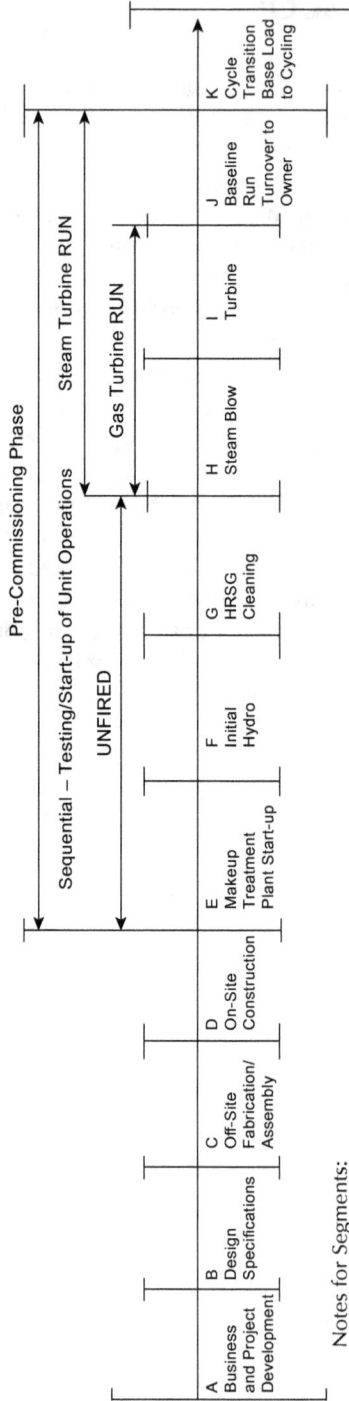

Notes for Segments:

A) Business/Project Development may take 6 months to 2 years to establish contracts and obtain permits.

B) Development of design specifications begins during segment A.

C/D) Construction can start as early as 6 months into the design specifications segment with land grading and initiation of support piles and structures. About 18 to 24 months total time is required for this segment.

E) Makeup water treatment plant startup will take approximately 1 week. Temporary piping and exhaust points for the steam blow are typically installed concurrently with this segment.

F) Initial hydrotesting and feedwater train cleaning and operational start-up requires about 4 days to 1 week.

G) Hydraulically flushing and chemically cleaning the heat recovery steam generator (HRSG) requires about 1 week, however up to 30 days has been reported.[1]

H) Steam blows, depending upon the number of flow paths chosen and the impingement target acceptance criteria, can take from 2 to 3 days.

I) Turbine start-up. The time, which includes dismantling the temporary piping and refitting the steam turbine (ST) to accept steam plus debugging turbine controls, will vary and may take 5 to 6 weeks.

J) Baseline run/performance testing will require 100 hours of continuous run. Total time will vary.

K) Cycle transition. Approximately 1 to 2 years are typically required to complete all the outstanding issues.

❑ Section 3 ❑
AVOIDING PREMATURE FAILURE

The potential for component failures may increase due to poor preplanning, improper selection and execution of design specifications, or inadequately performed pre-commissioning stages. The nature of these failures may be mechanical, operational, or chemical[9,10]. Maintaining and monitoring cleanliness during fabrication, shipping, site storage, field construction, and pre-commissioning cleaning will avoid lengthy start-up delays and post-commissioning failures.

Avoidance of premature failure is best realized by a cause-and-effect discussion regarding the most common failures experienced in a combined cycle plant. These failures may result from, but not be limited to, any of the following categories:

1. Construction issues, including fabrication, shop hydro, shipment, storage, and erection associated with:
   - Pitting corrosion
   - General corrosion
   - Microbiologically influenced corrosion (MIC)
   - Deposit formation
   - Contamination by chemicals and foreign materials

2. Preoperational issues, including hydro test, storage, chemical cleaning, steam blow, and lay-up associated with:
   - Pitting corrosion
   - General corrosion
   - Microbiologically influenced corrosion (MIC)
   - Deposit formation

3. Design and operational issues associated with:
   - Flow accelerated corrosion (FAC); both single- and two-phase flow.
   - Corrosion fatigue
   - Under-deposit corrosion
   - Hydrogen damage
   - Caustic gouging
   - Phosphate gouging
   - Pitting corrosion
   - Stress assisted corrosion
   - Thermal fatigue
   - Creep
   - High-cycle fatigue
   - Fretting corrosion
   - Crevice corrosion
   - Copper alloy related corrosion mechanisms (e.g. ammonia grooving or dealloying)
   - Microbiologically influenced corrosion (MIC)
   - Galvanic corrosion

Component failures associated with poorly planned or executed design decisions and pre-commissioning activities can have a long term impact or influence causing operational failures. Several examples are provided below to illustrate their importance.

- Poor pre-cleaning, drainage and preservation of fabricated components after shop hydrotesting (category 1 in Table 1) may result in localized or general corrosion that can continue during the pre-commissioning period (category 2 in Table 1). This may manifest itself after turnover to the plant owner as high-pressure evaporator tube failures due to pitting, under-deposit corrosion, stress assisted corrosion or, in a cycling plant, as corrosion fatigue.

- Porous corrosion deposits may develop during on-site storage and erection in uncapped or unsheltered tube sections exposed to high humidity or condensing moisture. Development of such corrosion can be avoided by excluding moisture, or by applying preservatives or vapor-phase corrosion inhibitors (VPCIs). Tube failures associated with under-deposit corrosion can result if deposits are not detected (e.g. by a video probe inspection) and removed by effective chemical cleaning prior to operation.

- Greater attention to design and materials selection is required for a cycling plant. Evaluation of the potential number of restarts per year, duration of dispatch, and the time available for return to service may justify the inclusion of additional equipment, such as an auxiliary boiler to minimize susceptibility to stress assisted corrosion. In addition, corrosion fatigue failures may develop if high ramp rates are used frequently during start-up[11,12]. Designing the system with all-ferrous metallurgy removes concerns associated with copper/copper alloy corrosion and permits optimization of corrosion control for ferrous materials by allowing operation at higher pH and dissolved oxygen concentrations thus minimizing the susceptibility to flow accelerated corrosion (FAC).

- The potential for crevice corrosion should be evaluated for all condensers with austenitic (300 series) stainless steel tubes. The Nickel Development Institute (NiDI) offers free software for evaluating the likelihood of crevice corrosion occurring in austenitic, duplex, and some other specialty stainless steels.[13] The most significant factors contributing to crevice corrosion include low molybdenum content, pH, sulfate concentration at high temperature, and chloride concentration. A main condenser used as a dump condenser can be particularly susceptible to crevice corrosion because of the exposure to high temperatures when steam is being dumped. The NiDI crevice corrosion model has been used to justify the installation of a more corrosion resistant alloy (i.e., a tube alloy containing a higher concentration of molybdenum), the installation of titanium tubing, and the implementation of more stringent water chemistry control limits.

Potential damage mechanisms and commonly associated locations are given in Table 1. Definitions of these mechanisms are included in Appendix A. When these mechanisms are initiated prior to commercial operation, degradation can be expected to continue when the plant is operational resulting in failure at a later date.

Table 1. Damage Mechanisms and Their Common Locations

Damage Mechanism	Category	Condenser	Feedwater Heater, Preheater, Low Temperature Economizer	Deaerator	Economizer	LP Evaporator	IP Evaporator	HP Evaporator	Super-heaters	Re-heaters	Steam Piping	Turbine	Rotor air coolers
General corrosion	1, 2, 3	x	x	x	x	x	x	x			x	x	
Caustic gouging	3						Rare	x	Rare				
Corrosion fatigue	3	Rare		x	x	x	x	x				x	
Stress assisted corrosion	3		x	x	x	x	x	x					
Creep; Creep fatigue	3								Rare	Rare	Rare	x	
Deposit formation	1, 2, 3	x	x	x	x	x	x	x	x	x	x	x	

Flow accelerated corrosion (FAC)	3	x	x	x	x	x	x					
Flow accelerated condensate corrosion	3	x										
High-cycle fatigue (Vibration)	3	x							Rare	Rare		x
High Temperature Oxidation	3								x	x	x	
Hydrogen damage	3							x				
Liquid Droplet Erosion	3											x
Microbiologically influenced corrosion (MIC)	[1, 2] 3	x	[x]	[x]	[x]	[x]	[x]	[x]	[x]	[x]		
Phosphate gouging	3							x				

Damage mechanism	Category												
Pitting corrosion (oxygen)	1, 2, 3	x	x	x	x	x	x	x			x	x	
Stress corrosion cracking (SCC)	3	x	x								x	x	x
Thermal fatigue	3								x	x			
Under-deposit corrosion	3	x					Rare	x					
Gas-side dew point corrosion	3		x										
Fretting corrosion	3	x											
Crevice corrosion	3	x									x	x	x
Galvanic corrosion	3	x											

Categories are defined as:
1 – Construction (fabrication, shop hydrotesting, shipment, storage, and erection)
2 – Pre-Operation (site hydrotesting, storage, chemical cleaning, steam blow, and lay-up)
3 – Operation

Damage mechanism terms are defined in Appendix A and the references where applicable

❒ Section 4 ❒

PROJECT DEVELOPMENT CONSIDERATIONS

There are many critical factors to consider when developing a project. However, many designers and constructors are contracted to build a plant that is already conceptually designed and permitted. These conceptual plant designs vary from project to project because of different site-specific factors. Therefore, it is impossible to define a template for typical project development.

Generally, the initial project development tasks include: a) choice of a suitable site with access to water, transmission lines, and fuel; b) definition of the type of operational duty required; and c) evaluation of project economics and permitting requirements. Project development considerations are illustrated in Figure 2. The size of the plant and whether it includes export steam with or without condensate returns must be defined. The heat recovery steam generator (HRSG), gas turbine, and steam turbine can then be configured appropriately.

The main decision drivers during project development are the type of plant, available water supply, environmental considerations, chemical regimes, and major equipment components (HRSG, condenser, steam turbine, and water treatment equipment).

❒ TYPE OF PLANT

Determining the way the plant will be operated early in the development of the project is important. Deciding whether a plant will normally be base-loaded or cycling is an important decision that can then be incorporated into the plant design. Base-loaded plants are designed to operate at full load, or at a steady base load, the majority of the time. Cycling units may operate anywhere from no load to full load at any time and are frequently shut down or bottled-up during the night hours and weekends.

❒ WATER SUPPLY[14]

A project cannot be developed without an adequate and reliable water supply that is also economically and environmentally viable. Several potential water sources may be available when developing, planning and executing power projects. There may be city water (potable water), surface water (river, lake, etc.), seawater, well water or gray water (secondary treatment effluent, etc.) or a blend therof. In addition, returned condensate from one or more host facilities may be fed back into the system. The quantity of returned condensate can vary from 0% to 90% of the total makeup water requirement, and have wide variations in temperature and purity.

Due to water supply restrictions, fresh water scarcity, and site permitting constraints, more plants are being forced to use non-traditional sources of water, such as gray water, production water, or mine drainage water, as the plant raw water supply. Gray water appears to be the most challenging source, in part due to the potential for sudden and unpredictable changes in quality, and because

Figure 2: Project Development Considerations

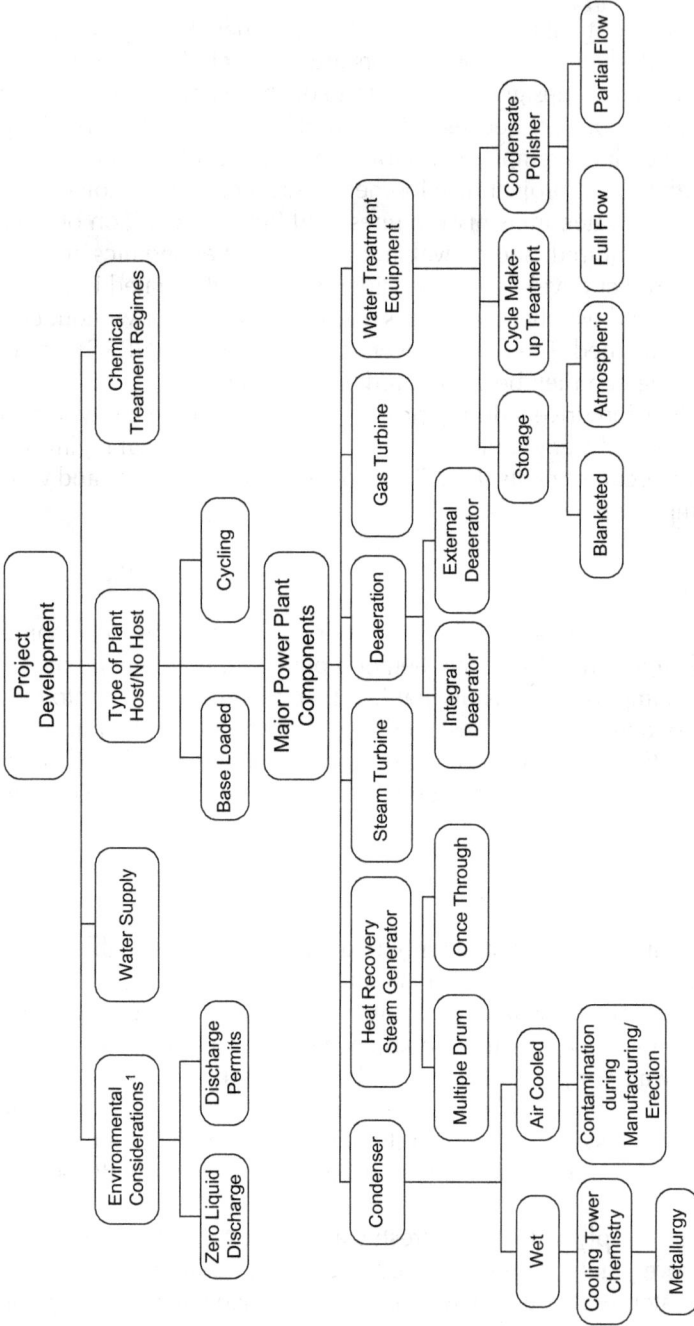

[1] Commonly omitted from the design is an auxiliary boiler to assist in maintaining layovers, lay-ups, and start-ups, and for Zero Liquid Discharge Plants

it can contain substantial amounts of COD, BOD, phosphates, nitrates, and ammonia. These contaminants are not usually seen at high concentrations in other water sources.

Chemical dosage and other design considerations are reasonably predictable for the traditional water sources such as well water or surface waters. While seasonal variations can have major impacts on most waters, it is the gray waters that provide the greatest challenges in terms of major variations in content.

❏ ENVIRONMENTAL CONSIDERATIONS

The National Pollutant Discharge Elimination System (NPDES) in the United States or equivalent federal agency permits for the plant can have a significant impact on the design requirements for the water systems and other plant equipment, especially, the plant's cooling water systems. Limits established in the discharge permit dictate how much of each chemical species can be discharged from the facility. These limitations can impact not only the chemical additives portion of the water treatment plan, but also the type of equipment and system metallurgy specifications. Therefore, it is imperative that the permit application be filed as early as possible so that the plant can be designed to meet the restrictions set forth in the permit.

In developing the application for the NPDES permit, the plant designer must investigate the various plant water balance options, plant operating modes, and potential chemical treatment schemes to determine how the different approaches will affect the plant's wastewater discharge profile. As an example, operating with multiple cycles of concentration in the cooling tower will increase the concentration of contaminants in the blowdown stream, but will decrease the volume of wastewater discharged. Depending on the concerns associated with the receiving body of water, either scenario may be preferred by the regulating authorities.

During the planning phase, the developer also needs to work with the water treatment chemical supplier(s) to ensure that all treatment chemicals that the plant might use are included in the initial NPDES permit application. If cleaning waste is to be discharged to the site waste treatment facility, this information must also be included in the discharge permit. Otherwise, off-site disposal will be required. It is much easier to decide not to use a chemical after it has been permitted than to add another chemical to the treatment plan after the NPDES permit has been issued.

NPDES permits are not the only environmental permits that the developer has to consider when developing the plant water balance. The frequency with which plant air permits impact the plant water balance and the way that the systems are engineered is increasing with time. The United States National Ambient Air Quality Standards (NAAQS) or equivalent federal agency set restrictions on particulate matter (PM). In certain areas this limit can directly or indirectly set restrictions on cooling tower emissions and plant cooling water system operations. Federal agencies frequently place restrictions on PM less than or equal to 10 micrometers (PM_{10}), 5 micrometers (PM_5), or 2.5 micrometers ($PM_{2.5}$). The PM limit can restrict the number of cycles of concentration permitted during operation of the cooling tower by limiting the allowable total dissolved solids (TDS) concentration in the cooling tower drift and, thus, in the bulk circulating water. The maximum allowable TDS of the circulating water can be calculated by the following equation:

$$\text{Maximum Allowable TDS (ppm)} = \frac{(PM \times 1{,}000{,}000)}{(Q \times D \times 1440 \times 8.34)}$$

where: PM = Permit limit (lbs./day)
Q = Circulating water flow rate (gal/min)
D = % Drift rate (as a decimal, %/100)

❏ MAJOR EQUIPMENT COMPONENTS

Decisions made during the project development phase play an important role in determining the options available when equipment design is specified. The system components and their intended service duty are determined during project development. The allowable operating limits for each piece of generating equipment must be evaluated to verify operation within design parameters such that warranty coverage will not be voided. These same operating limits should be used to develop the design specifications and operating parameters associated with ancillary equipment.

Makeup water treatment systems typically consist of two subsystems: a pretreatment system followed by a demineralization system. Selection of the pretreatment system equipment depends on the quality and source of raw water and the equipment specified for the demineralization system. The demineralization system further purifies the makeup water to meet the purity requirements for the HRSG feedwater and the associated steam turbine.

❏ CHEMICAL REGIMES[15,16]

Selection of the chemical treatment should take into consideration plant equipment (e.g. turbine requirements), anticipated plant operating modes, personnel staffing, and performance requirements. The evaporators must all be protected from deposit accumulations and the low pressure (LP) and intermediate pressure (IP) evaporators, in particular, from FAC. The high pressure (HP) evaporator should be protected from damage due to trace levels of feedwater contamination and; the steam turbine requires protection from damaging contaminant intrusion.

Feedwater and condensate systems must be protected from oxygen attack and flow accelerated corrosion (FAC).

☐ Section 5 ☐

PROJECT DESIGN SPECIFICATIONS

☐ CRITICAL DESIGN PARAMETERS

Critical parameters for the design of a combined cycle power plant are itemized in Table 2. Additional details are included later in this section.

☐ TYPE OF PLANT

Base-loaded plants are designed to operate at full load, or at a steady base load, the majority of the time. Cycling units may operate anywhere from no load to full load at any time and are frequently shut down or bottled-up during the night hours and weekends.

If the plant is to operate routinely in a cycling mode, the anticipated frequency and duration of the cycles must be addressed in the plant design phase. Under cycling conditions, the expansion and contraction rates of critical components such as hangers, joints, piping, duct work, coils, tubes, and tube-to-tube header connections must be included in the design to avoid thermal fatigue failures. Specifying a design to include cycling will affect start-up ramp rates and possibly the details of construction of the equipment. Some of the items that must be designed to accommodate operation not only for normal or base-loaded operation conditions, but also for frequent start-up operations, include the chemical feed system, makeup water treatment system, HRSG blowdown system, drain line locations, drain pots for condensation with conductivity sensors, stack dampers, and insulation. When the deaerator and/or steam turbine condenser are not in service, auxiliary steam availability and a correctly sized steam sparger in the demineralized water storage tank and condenser hot well should be present to provide warm start-up conditions and some deoxygenation.

If the plant is to supply steam to an outside host, additional preplanning during the project design and development phase can minimize problems during plant start-up and operation. Items that should be addressed in the design of a cogeneration facility include host steam chemistry limitations, collection and treatment of the returned condensate, chemistry limits for acceptance of returned condensate, sizing of the makeup treatment system to replace dumped condensate, disposition of dumped condensate, and operation without steam export.

Table 2. Critical Design Parameters

1. Steam host or stand-alone
   a) Requirements for and restrictions on host steam purity
   b) Condensate - amount and quality recovered vs. returned purity required

2. Climate/weather/location of plant
   a) Equipment redundancy
   b) Freeze protection
   c) Shipping/procurement time for equipment and chemicals

3. Gross rating
   a) Power augmentation that will be applied
   b) Secondary fuel type that will be used

4. Operation mode; base-load or cycling
   a) Manpower requirement
      1. Degree of automation
      2. Instrumentation
      3. Maintenance requirements
   b) Ramp rates allowed (environmental – air permit)

5. Water source and history
   a) Amount required vs. availability
   b) Chemistry
   c) Review of possible treatment options
      1. Reliability
      2. Life cycle (maintenance/replacement costs)
   d) Amount and quality of condensate from host

6. Water balance

7. Heat balance
   a) Condenser cooling water supply

8. Discharge permit or zero liquid discharge

9. Cycle chemistry desired, required, and selected
   a) Customer preferences/requirements
   b) Condensate polishing required? Type selected?

10. Materials selection for equipment and piping

11. Major equipment specification
    a) Gas and steam turbines
    b) Condenser - cooling tower, air-cooled, or wet-surface air-cooled
    c) HRSG
    d) Feed pumps

12. Auxiliary equipment specification
    a) Deaerator
    b) Makeup/pretreatment
    c) Sampling/monitoring panels
    d) Condensate polishing
    e) Auxiliary boiler/steam supply
    f) Chemical feed equipment

13. Cleanliness of major equipment fabrication
    a) Hydro testing, cleaning and passivation requirements
    b) Inspection (NDE)
    c) Preservation for shop storage and transportation

14. Site storage/construction
    a) Cleanliness
    b) Preservation of fabricated equipment and components
    c) Size and installation of hydro testing drains, connections for chemical feed, connections for chemical cleaning, and temporary recirculation system steam lines and mufflers for steam blows

15. Pre-commissioning/Start-up
    a) Plan adequate time and water resources for pre-commissioning and start-up activity

☐ WATER SUPPLY[14]

Once the water source is established, developing the design basis water analysis is the next major challenge. Average and worst case percentage makeup and percentage condensate return to the HRSG feedwater should be assessed. Collection of multiple samples of the raw water and host condensate return if present should be completed over an extended period of time to capture seasonal and other variations in water quality. The parameters prescribed for the different water sources are summarized in Table 3. The sampling program should continue throughout project execution to verify quality consistency and minimize "surprises".

When possible, a more practical approach for determining the variations in a water source is to survey users of the same water and factor their experience into the design. Experience from other users, if properly applied, can forestall many problems not captured by the design basis water analysis. Historical analysis records are a valuable resource and should be considered if available.

While it is impossible to provide a guideline for water sampling that would serve all plant types and all water sources, a general guide that provides a good representation of the parameters to be included is as follows:

• For well water supplies, samples should be collected from each well.
• Surface water supplies can be much more variable and require more samples to characterize the water supply. Ideally, there will be a source of data (government monitoring agency, etc.) indicating the annual fluctuations of water quality for a few parameters. This should be supplemented with at least 10 water analyses, ideally spread over at least 12 months.

Table 3.   Supply Water Analyses

REQUIRED TESTS (ALL WATER SUPPLIES)

Cations
- Ammonia
- Barium
- Calcium
- Magnesium
- Potassium
- Sodium
- Strontium

Anions
- Bicarbonate (from M alkalinity)*
- Bromide
- Carbonate (from P and M alkalinity)*
- Chloride
- Fluoride
- Nitrate
- Nitrite
- Phosphate (ortho and poly)
- Organic acids (for cycle makeup treatment system influent)
- Reactive silica
- Sulfate

Metals
- Aluminum
- Iron (dissolved and total)
- Manganese (dissolved and total)

General Parameters, Foulants, and Organic Matter
- Silt Density Index (SDI) (for reverse osmosis systems)
- Color
- Colloidal silica (Determined by subtracting reactive from total silica)
- Conductivity
- pH*
- Total organic carbon (TOC), a.k.a. Non-purgeable organic carbon (NPOC)
- Total silica (by inductively coupled plasma)
- Total suspended solids (TSS)
- Turbidity

SUPPLEMENTAL TESTS BASED ON SITE-SPECIFIC REQUIREMENTS

Anions
- Organic acids (non-cycle make-up waters)
- Sulfide (for surface and well waters)
- Boron/borate

Dissolved Gases
- Methane
- Carbon dioxide*
- Hydrogen sulfide

Metals
Heavy metals (e.g., arsenic, copper, iron, lead, zinc, vanadium, mercury, selenium, etc.) Some are part of the priority pollutants

General Parameters, Organic Matter And Foulants
• Chlorine demand*
• Oil & grease (O&G)
• Phenols
• Total dissolved solids (TDS)
• Volatile organic carbon (VOC)

Bacteriological Tests	
• BOD$_5$	• Sulfur-reducing bacteria
• COD	• Total heterotrophic bacteria and anaerobic bacteria plate counts
• Iron-related bacteria	

Miscellaneous
• Priority Pollutants**

* pH, alkalinity, and temperature require on-site analyses for accuracy.
** Refer to US Clean Water Act Federal Register List 40CFR 401.15 for the list of toxic or priority pollutants.

- For city or municipal water supplies, sometimes all of the necessary data will be available from the water supplier. However, testing for suspended matter and heavy metals is suggested, since additional material can be contributed by potable water service lines.

Table 3 is separated into two parts. The first shows the parameters required for all water supplies, and the second shows parameters that are less commonly required. Individualized lists of parameters and minimum sampling requirements should be provided in consultation with a water treatment specialist.

❏ WATER TREATMENT EQUIPMENT

Makeup water treatment systems typically consist of two subsystems: a pretreatment system followed by a demineralization system. The pretreatment system removes materials such as suspended solids, bacteria, color, iron, manganese, aluminum, phosphate, and oil from the water to prevent fouling of the demineralizer system that follows. The demineralization system removes dissolved solids and some dissolved gases (e.g., carbon dioxide) to meet the purity requirements for the HRSG feedwater and the associated steam turbine. Selection of the pretreatment system equipment depends on the quality and source of raw water and the equipment specified for the demineralization system.

PRETREATMENT SYSTEM

Water supplied to the makeup water treatment system may be surface water, from a lake or river; well water; municipal potable water; or treated wastewater. Desalinated water has not been included as a potential water source due to the additional complexity and cost of such a system.

Surface waters usually require pretreatment. Chlorination, followed by conventional clarification and filtration, is typically used. However, use of an ultrafiltration system in combination with chlorination and dechlorination may be sufficient depending on the quality of the surface water supply.

Well water usually contains low concentrations of suspended matter and bacteria but can contain substantial amounts of iron, manganese, and dissolved gases such as carbon dioxide and hydrogen sulfide. For well water sources with low iron and manganese content, pretreatment usually involves single stage filtration before the demineralization process. Greensand filters may sometimes be required to remove high levels of iron and manganese present in well water. Contaminated well water has been excluded from this discussion due to the variability and complexity of this as a water source.

Municipal potable water is essentially pretreated and usually requires only the removal of trace amount of suspended solids and oxidants used as disinfectants such as chlorine or chloramines. This removal can be achieved with an activated carbon filter or a media filter followed by the injection of a reducing agent such as sulfite.

Treated wastewater may originate from a secondary wastewater treatment plant with further disinfection in a tertiary waste treatment plant which typically contains some type of media filtration. There will be residual ammonia, phosphorus, TOC (COD), and low concentrations of iron present in such treated effluent. Therefore, clarification followed by multimedia filtration or ultrafiltration should be included in the pretreatment process prior to demineralization. The hardness and alkalinity will probably be lower than those of surface and ground waters found in the area.

For ion-exchange type demineralization systems, the treatment discussed above will suffice. For membrane-based make-up demineralization systems, an ultrafiltration system rather than granular media filtration is typically selected for pretreatment. Two-stage filtration is recommended ahead of an RO system which should always be preceded by cartridge filtration regardless of pretreatment provided upstream.

Selecting the proper size of the water treatment equipment and the correct combination of unit operations is essential for both the short-term and long-term operation of the plant. It is important to perform a thorough review of water quantity and quality requirements of the different users. A preliminary design should be developed to determine the best and most cost-effective system. Final selection should be made with the help of an experienced water treatment specialist. While this is not intended to be a selection guide, some of the salient items are presented below.

The important issues to be considered are as follows:

- Source(s) of water and the different combinations
- Reliable water data, including seasonal variations
- Purity or quality requirements of all consumers
- Experience by owner or neighboring plants using the water(s) to be treated
- Operating costs (labor, power, chemicals, etc.)
- Methods of makeup treatment (resinous ion exchange, membrane UF/RO/ EDI/EDR, evaporation)

- Consideration for initial removal, degasification, and exclusion of oxygen prior to the steam cycle and after the makeup water treatment system
- Option to include condensate polishers (types, flow rates, etc.)
- Cooling tower makeup and/or side-stream treatment of dissolved and suspended solids
- Wastewater generation (quantities and qualities) for conformance with discharge permit limits
- Wastewater treatment and recycling options

Provisions should be made for the use of mobile pretreatment equipment to handle initial commissioning and emergency upsets. This requirement may be accomplished simply by adding flange connections and locating an electrical power supply connection for mobile equipment.

Various water pretreatment methods are provided in Table 4.

DEMINERALIZATION

Typically, two-stage demineralization is required to achieve the level of feedwater purity required for high-pressure (>1000 psig or > 6.9 MPa) HRSG units. The first stage is referred to as primary demineralization and the second stage as secondary (or polishing) demineralization. The primary demineralization system typically consists of cation and anion exchangers or reverse osmosis units. The secondary demineralization system typically consists of a mixed-bed exchanger or electrodeionization (EDI) unit. Most cation and anion demineralizers are counter-currently regenerated deep bed demineralizers; however, co-currently regenerated deep bed demineralizers do exist at some facilities.

Common demineralization system designs are presented in the following list.

1. Reverse osmosis (RO) followed by membrane degasification and a continuous EDI unit.
   - This combination of membrane units has the benefit of not needing chemical regeneration, but chemicals are needed to clean the RO and to control corrosion, scaling, and biofouling.
2. A double-pass RO followed by a degasifier and EDI or mixed bed.
   - This combination is used for water with very high dissolved solids content or for production of higher purity effluent.
3. RO followed by a cation exchanger, decarbonator or degasifier, and anion exchanger.
   - The decarbonator is not needed for water supplies with low alkalinity. A mixed bed polisher may be required downstream of the anion exchanger.
4. Two bed demineralizers, sometimes with an intermediate decarbonator, followed by a mixed bed polisher.
   - This arrangement is one of the more traditional makeup treatment systems.

Many other systems combinations are feasible but the preceding list contains those most commonly installed for the makeup water treatment systems utilizing existing technologies.

Table 4. Water Pretreatment Methods

Impurity	Methods of Removal	Typical Residual After Treatment[1]
Alkalinity	a. Cold lime precipitation[2] b. Warm and hot lime precipitation[2] c. Chloride anion ion exchange d. Weak acid cation ion exchange e. Reverse Osmosis f. Strong acid cation	a. > 35 ppm as $CaCO_3$ b. >17 ppm as $CaCO_3$ c. >8 ppm as $CaCO_3$ (10% of influent M alkalinity) d. If TH>M alkalinity, approx. 5-10 ppm as $CaCO_3$. If TH<M, depends on inlet analysis[3] e. 1-5% of influent f. Below detection limit
Ammonia	a. Gas transfer membrane b. Steam stripping c. Vacuum degasification d. Strong acid cation ion exchange	a. 95% removal and pH dependent b. pH dependent c. pH dependent d. 0.01 – 2 ppm
Carbon dioxide, free	a. Atmospheric decarbonation b. Gas transfer membrane c. Vacuum degasifier (aka "vacuum degasser", "vacuum deaerator") d. Cold, warm and hot lime softening[2]	a. 5 - 10 ppm as CO_2 b. 2 – 10 ppm as CO_2 c. 0.01 ppm as CO_2. Below detection limit
Chloride	a. Ion exchange demineralization b. Reverse osmosis	a. 0.003 - 3 ppm as $CaCO_3$ b. < 1 - 5% of influent
Color and Organic (Removal efficiency dependent on type of organic)	a. Clarification/media filtration b. UF[4] c. Chemical oxidation[5]	a. 5 - 10 APHA color b. 1 - 5 APHA color c. 5-10 APHA color

Contaminant	Treatment Methods	Values
Colloidal silica Colloidal iron	a. Chemical feed, clarification + filtration b. Polyelectrolyte feed + filtration c. Ultrafiltration	a. Dependent on analysis b. Dependent on analysis c. Dependent on MWCO[6]
Fluoride	a. Cold lime, warm lime, hot lime softening[2] b. Ion exchange demineralization c. Reverse osmosis	a. 10 ppm as F b. < 1.0 ppm as $CaCO_3$ c. < 1 – 5% of influent
Hardness	a. Cold lime-soda precipitation[2] b. Warm lime and hot lime softening[2] c. Sodium cycle cation softening d. Ion exchange demineralization e. Single pass reverse osmosis f. Second pass reverse osmosis after primary single (first) pass reverse osmosis	a. < 34 - 85 ppm as $CaCO_3$ b. < 17 - 25 ppm as $CaCO_3$ c. < 1.0 ppm as $CaCO_3$ d. << 0.2 -1.0 ppm as $CaCO_3$ e. < 1 - 5% residual depending on water supply and membrane f. <<1.0 ppm as $CaCO_3$
Hydrogen sulfide	Aeration Chlorination	<1 ppm as H_2S
Iron and Manganese	a. Air injection + filtration - low iron and CO_2 with high alkalinity b. Aeration + filtration - moderate iron, high CO_2 c. Aeration, chemical feed, clarification + filtration - high iron and manganese with organics present d. Polyelectrolyte feed + filtration e. Greensand filter	a. <0.3 - 0.5 ppm as Fe or Mn b. <0.3 - 0.5 ppm as Fe or Mn c. <0.5 ppm as Fe or Mn d. <0.1 - 0.3 ppm e. <0.5 ppm as Fe or Mn Note, Mn in effluent usually <10% of iron

Methane	a. Aeration b. Steam stripping	a. < 1 ppm as CH_4 b. < 1 ppm as CH_4
Oxygen	a. Deaeration - Pressure b. Deaeration - Vacuum c. Gas transfer membrane	a. 0.007 ppm as O_2 b. 0.05 - 0.3 ppm as O_2 c. ≤0.01 - 0.02 ppm as O_2
Phosphate, Orthophosphate and Polyphosphate	a. Reverse osmosis b. Cold lime precipitation[2] c. Hot lime precipitation[2]	a. < 1 - 5% b. < 1 ppm c. < 1 ppm
Silica (reactive)	a. Chemical feed, coagulation + filtration - cold temperatures b. Warm and hot lime precipitation[2] with sufficient magnesium present c. Ion exchange demineralization d. Reverse osmosis e. Chemical feed, clarification + filtration f. Polyelectrolyte feed + filtration	a. 2 - 3 ppm, Mg and pH dependent as SiO_2 b. 0.5 - 1.0 ppm as SiO_2 c. < 0.01 – 0.1 ppm as SiO_2 d. ≤0.01 - 0.10 ppm as SiO_2 e. Dependent on analysis f. Dependent on analysis
Sodium	a. Ion exchange demineralization b. Primary Single pass reverse osmosis c. Second pass reverse osmosis after single (first) pass reverse osmosis	a. 0.003 – 2 ppm as $CaCO_3$ b. < 1 - 5 % residual depending on water supply and membrane. c. <1.0 ppm as $CaCO_3$
Sulfate	a. Barium precipitation b. Ion exchange demineralization c. Reverse osmosis	a. 17 - 25 ppm as $CaCO_3$ b. <0.003 ppm as $CaCO_3$ c. 1 - 5% of influent

Turbidity (Removal efficiency dependent on type of turbidity)	Filtration – a. sand alone b. anthracite and sand dual media c. anthracite, sand and garnet multimedia d. Chemical feed and filtration e. Chemical feed , clarification and filtration f. Ultrafiltration or microfiltration (UF/MF)	a. < 2 NTU, >90% removal at >25 micron b. < 2 NTU, >90% removal at >25 micron c. <2 NTU, >90% removal at >15 micron d. < 1.0 NTU e. < 1.0 NTU f. << 1.0 NTU

Notes:

1. The typical "Residual after Treatment" achieved depends upon the type of equipment selected, equipment configuration and staging arrangement. It is affected by operating temperature and hydraulics, as well as the selected reagents and regenerants and their application concentrations.

2. In cold, warm (>140° F) and hot (>212° F) lime softening, removal of all species generally improves with temperature.

3. TH = Total Hardness, M = Total Alkalinity

4. Membrane systems remove organics and color dependent on MWCO.

5. Where "Chemical Oxidation" is noted, the process may use chlorine, chlorine dioxide, ozone, peroxide, ultraviolet light and peroxide, or ozone and peroxide combinations.

6. MWCO = Molecular weight cut-off

EQUIPMENT SIZE AND EFFLUENT QUALITY

The pretreatment system at a combined cycle plant may be designed with a much larger capacity than the demineralization system to provide pretreated water for other uses, such as cooling tower makeup, service water, and auxiliary cooling water and, at some facilities, potable water.

The demineralized water makeup rate can vary from 1-5% of the steaming capacity for stand-alone combined cycle power plants to over 50% for cogeneration facilities. For multi-stage demineralization systems, the typical demineralized effluent quality should be <10 ppb of silica as SiO_2, <0.1 $\mu S/cm$ specific conductivity at 25°C, and <200 ppb of total organic carbon.

Makeup water quality parameters are established to ensure compliance with main and reheat steam purity limits. Other limits are then back-calculated to determine saturated steam purity limits, HRSG water limits, feedwater limits, condensate limits, and makeup water quality limits.

- The saturated steam purity limits can be estimated based on the main steam limits and the attemperation rates (% of feedwater sprayed into steam).
- The HRSG water limits[17] generally are determined by the operating pressure and type of internal chemical treatment (all-volatile treatment, phosphate treatment, etc.).
- Feedwater quality is determined by the HRSG drum water as well as the HRSG drum blowdown rate. The makeup water quality requirements are based on HRSG feedwater quality and condensate return.

Consider purchasing pre-regenerated ion exchange resins and using pharmaceutical grade resins with very low leachable TOC. If there already is a host end user, utilize historical makeup and condensate return data to improve makeup water treatment equipment selection, sizing, and regeneration sequencing[17, 18].

It should be noted that most of the common epoxy coatings utilized to protect the interior surfaces of demineralized water storage tanks have a maximum service temperature around 250°F (121°C) and are subject to degradation from contact with steam. Therefore, it is recommended that the demineralized water storage tank or condensate storage tank be constructed of stainless steel if steam sparging is to be utilized.

❑ CONDENSATE POLISHERS[20]

Condensate polishers are ion exchange resin-based systems used to remove contaminants from the HRSG feedwater cycle. Typically, they are installed near the discharge of the condensate pumps to remove dissolved contaminants, such as sodium and silica, as well as suspended material, such as particulate iron, from the condensate/feedwater and to assist the plant in meeting the feedwater and steam purity requirements of HRSG and turbine suppliers. There are several plant design situations in which condensate polishers should be included as part of the condensate/feedwater system design. These are:

- Plants where brackish water or seawater is used to cool the main condenser

- Plants where condensate from exported steam is returned to a steam cycle operating above 1200 psig (8.27 MPa)
- Plants with a steam cation conductivity limit ≤ 0.15 μS/cm

Condensate polishers are also recommended for plants that are designed with two or more of these six design and operating characteristics which have the potential to impact plant performance and start-up:

- Plants with steam cation conductivity limits ≤ 0.20 μS/cm
- Plants where an air-cooled condenser (ACC) is utilized to cool steam turbine exhaust
- Plants that operate with an all-volatile treatment (AVT) or oxygenated treatment (OT) chemistry program
- Plants with HP steam operating pressure greater than 2400 psig (16.55 MPa)
- Cycling units with short start-up times
- Plants with a steam turbine manufacturer limit on LP steam cation conductivity included in contract

Even if the plant does not operate with any of these waters or conditions, condensate polishers may still be a wise investment. Polishers can reduce the time required to obtain steam purity during plant start-up, allow the unit to operate for a limited time with a condenser leak, allow for an orderly shutdown of the unit if a condenser leak occurs, reduce the requirement for demineralized water makeup by decreasing steam cycle blowdown requirements, and provide savings associated with the cost of demineralized water.

If condensate polishing equipment is not installed, rental ion exchange systems, designed as part of a sidestream clean-up loop, can be used after the steam blows to facilitate faster clean-up. Connections for this equipment must be planned for in advance.

☐ STEAM AND GAS TURBINES

Specifications for steam turbines are typically performance-based. Therefore, it is essential to coordinate and lock in the plant design water and heat balances for the steam turbine, gas turbine, and HRSG suppliers as early in the project development phase as possible. A single plant design water/heat balance is not sufficient. Multiple potential operating conditions need to be evaluated based on different anticipated plant scenarios such as fogging, power augmentation, and attemperation. In particular, if equipped with duct burners, the plant heat balance during duct burner operation needs to be addressed. The worst case scenario should then be used to back-calculate the water quality required to meet the specified operating limits for sodium, potassium, silica, chloride, sulfate, phosphate, and total organic carbon

For steam purity, this calculation will include the impurities in the drum water that are transported by both mechanical carryover (pressure dependency chart) and vaporous carryover (impurity distribution ratio at various pressures), plus the contribution of any attemperation water. Based on these calculations, the makeup treatment and condensate polishing equipment can be configured and sized. Operational control ranges can then be established for water conditioning equipment, drum chemistry, and steam purity. With this information in place, the preferred water treatment chemistry regime can be selected, taking into account operational flexibility and provisions for modification during incidents such as condenser in-leakage, condensate contamination, or chemical pretreatment upsets[1,17,21,22]. Refer to Figure 3 for additional guidance[17].

Since steam purity requirements vary not only with turbine operating temperature and pressure, but also with turbine manufacturer, the plant designer should obtain these limits from the manufacturer for review during the bid stage. Turbine steam purity limits are the basis for most other steam cycle-related equipment design, selection, and operational control limits[1,24,25,26]. Although pH is not measured, this limit is specified to minimize steam generator and piping corrosion. When the gas turbine design includes steam injection or cooling, the steam turbine steam purity requirements need to be compared to the steam purity requirements for the gas turbine. Then, utilizing the more restrictive requirements, the plant designer should determine the required makeup treatment system water quality parameters and the permissible cycle chemical treatment regimes.

Matching gas turbine and steam turbine cycles during transients and low load conditions is difficult. The greatest difficulty is experienced during start-ups, shutdowns, and steam turbine trips. Because the gas turbine produces so much exhaust heat at such a rapid rate the thermal ramp rate limits, as well as the permissible metal-to-steam temperature gradients (i.e., $\Delta T = 145\text{-}200°F/\text{hour}$ or $63\text{-}93°C/\text{hour}$), are often violated in thick-walled components of the HRSG and/or steam turbine during these transients.

Controlling these variables can require special gas turbine firing controls, a flue gas by-pass stack and damper, and/or a steam turbine by-pass system/ condenser dump. The latter can be accomplished in either a cascading or parallel fashion, but the former is more desirable. Sizing the dump system typically requires consideration that 140% of the steam turbine's 100% load is ejected during a trip[27]. Therefore, sizing of valves, piping, and dump tubes (spargers in the condenser) is very important, as is control of attemperation of the high-pressure by-pass steam and reheat by-pass steam.

Feed-forward schemes are typically used where the pressure and temperature upstream of the PCV (pressure control valve) and pressure downstream of the PCV are measured. The attemperation water flow is then based upon the enthalpy and mass flow rate of the dump steam during the transient, decreasing component failures associated with excessive superheat steam temperature. Erosion and vibration are handled by line/valve/sparger sizing and sparger system location and configuration within the condenser (i.e., full width of the condenser, with sparger tubes perpendicular to the condenser tubes and the sparger jets from the sparger tube parallel to the condenser tubes).

Preliminary copies of the steam turbine operating and maintenance (O&M) manuals and the erection manuals should be requested from the manufacturer at least several weeks prior to delivery of the equipment to allow time for preplanning the installation sequence and the coordination of labor and equipment availability.

Various preservatives may be applied to the steam turbine parts in the shop prior to shipment, including waxes, asphalt-based coatings and corrosion-

Figure 3:   Chemical Treatment Regimes

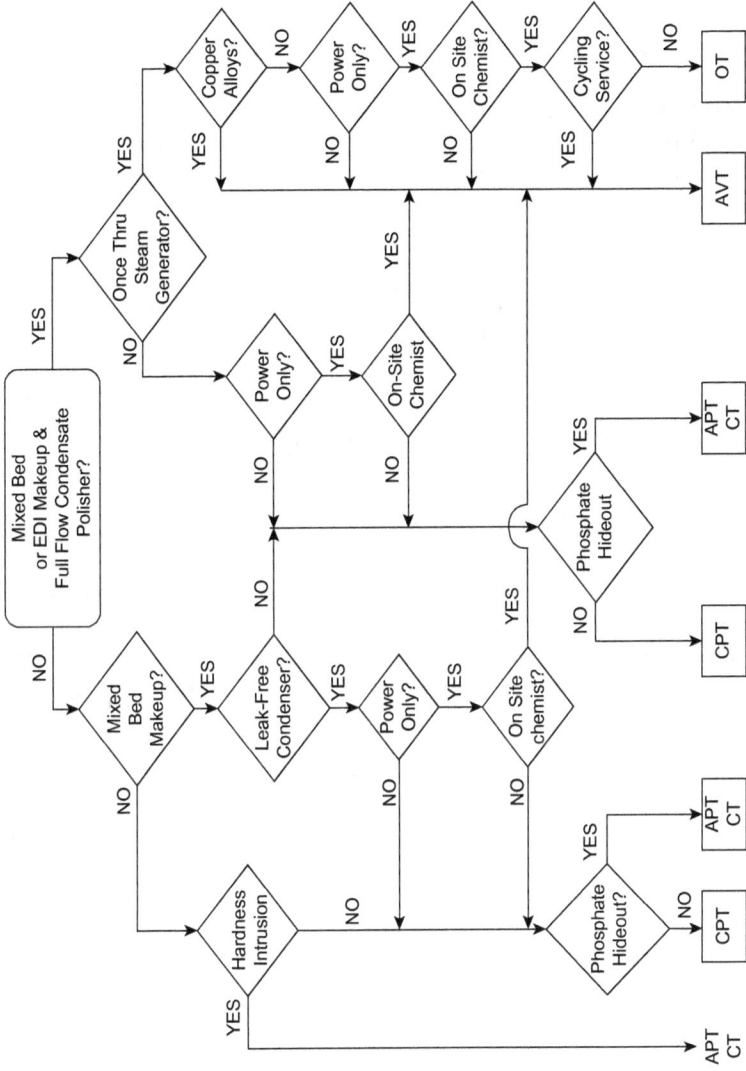

inhibiting oils. The anticipated length of storage time prior to installation or start-up of the unit may determine which preservatives are applied. Coordinating a "just-in-time delivery" of the turbine can minimize the number of parts coated with preservatives other than oil, and can decrease the volume of potentially hazardous waste that needs to be handled and disposed of at the site.

❒ CHEMICAL REGIMES[16,17]

Selection of the chemical treatment should take into consideration plant equipment (e.g. turbine requirements), anticipated plant operating modes, personnel staffing, and performance requirements. The evaporators must all be protected from deposit accumulations and the low pressure (LP) and intermediate pressure (IP) evaporators, in particular, from FAC. The high pressure (HP) evaporator needs to be protected from damage due to trace levels of feedwater contamination and; the steam turbine needs to be protected from damaging contaminant intrusion.

Feedwater and condensate systems must be protected from oxygen attack and flow accelerated corrosion (FAC). The pH of the feedwater, condensate, and LP evaporator needs to be increased by adding a volatile alkaline chemical such as ammonia and/or an organic amine to control FAC. When selecting a volatile alkaline chemical or oxygen scavenger, the ability of the chemical to protect all parts of the system, the effect the chemical will have on steam cation conductivity, and any concerns related to Occupational Safety and Health Administration (OSHA), Food and Drug Administration (FDA), and other applicable requirements/regulations should be considered.

In most HRSG systems the feedwater treatment regime of choice is all volatile treatment (AVT). In this treatment, only volatile alkalizing chemicals such as ammonia and neutralizing amines are used for pH control, potentially in conjunction with volatile oxygen scavengers. There are two regimes of AVT treatment:

- AVT(R) -a "reducing" environment chemistry that uses mechanical deaeration in combination with a chemical oxygen scavenger to maintain a relatively negative oxidation reduction potential (ORP) environment
- AVT(O) -a slightly oxidizing chemistry that uses only mechanical deaeration to maintain 5-10 ppb dissolved oxygen (DO) in the feedwater entering the economizer to create a slightly positive ORP environment

For plants with copper alloys present in the feedwater system, AVT(R) is recommended, as oxygen concentrations greater than 5 ppb can lead to increased corrosion of these alloys, especially in systems using ammonia for pH control. Many new systems designed strictly for power generation are constructed without copper alloys. When high-purity feedwater is maintained in these plants, the minimization of dissolved oxygen during operation is not critical and concentrations of 5 -10 ppb DO at the economizer inlet may prove beneficial in minimizing single-phase FAC. However, when a significant concentration of oxygen is present in the condensate or feedwater, any contaminant ingress can lead to rapid corrosion and equipment failure. Plant operation under those conditions should not be allowed.

Consequently, oxygen scavengers are recommended, i.e. AVT(R), for systems with copper alloys, systems in cycling service, systems with steam hosts, and systems where adequate chemical supervision is not available to respond to unexpected upsets in feedwater purity. The required feedwater purity for these AVT chemistries is typically <0.2 μS/cm cation conductivity.

Oxygenated treatment (OT) can also be an option for both once though type HRSGs and drum type systems that are constructed with all steel feedwater system metallurgy and operate with a feedwater purity limit of <0.15 μS/cm. For once through type HRSGs, feedwater DO concentrations are typically maintained in the range of 30-150 ppb with a pH of 8.5-9. For drum HRSG systems, feedwater DO control limits are set to maintain a maximum HRSG drum water DO concentration of 5-10 ppb based on a downcomer sample point.

Phosphate treatment is most often used when HRSGs send steam to and receive condensate back from a host plant.

The process of selecting the chemical treatment program for the HP evaporator should include consideration of many plant parameters including: steam purity requirements, type of makeup water treatment, presence and type of condensate polishers, potential for condensate contamination, condensate and feedwater system metallurgy, HRSG design, operating pressure, the use of duct burners, cycling service and the availability of a chemist to quickly implement corrective action should it be required. A decision tree for chemical treatment selection is illustrated in Figure 3 which presents some of the factors to be considered when selecting the HP evaporator chemical treatment regime.

The chemical treatment regimes shown in Figure 3 are as follows:

- All-Volatile Treatment (AVT)
- Alkaline phosphate treatment (APT)
- Caustic Treatment (CT)
- Congruent Phosphate Treatment (CPT)
- Oxygenated Treatment (OT)

Once all water conditioning equipment and chemical treatment programs are selected and test control ranges and action levels established, analytical testing and monitoring equipment can be selected, as well as the sampling locations and sample conditioning methods[21,22,23,25,26,]. Guidance on operational chemistry control limits can be found in ASME publication, "Consensus on Operating Practices for the Control of Water and Steam Chemistry in Combined Cycle and Cogeneration Power Plants"[17]. The associated sampling and monitoring required for chemistry control can be found in ASME publication, "Consensus on Operating Practices for the Sampling and Monitoring of Feedwater and Boiler Water Chemistry in Modern Industrial Boilers"[24]. The design specifications and baseline calculations found in the above referenced ASME publications are used in the preplanning for both the pre-commissioning and the commissioning phases of the combined cycle plant.

❒ HEAT RECOVERY STEAM GENERATOR

Stand-alone deaerators typically provide additional flexibility in selection of the chemical treatment program for the LP evaporator. The use of an integral deaerator limits the chemical treatment program to AVT and requires specifying upgraded materials for this section of the HRSG to prevent FAC issues. If the LP drum is utilized as a feedwater storage tank or serves as the source of attemperation water, however, AVT chemistry must be used as the treatment program for the LP section.

Additionally, the system metallurgy should ideally be copper-free since this increases chemical treatment options for the system.

Specifications for the HRSG should define:

- Location where the HRSG will be fabricated
- Degreasing of the HRSG by the fabricator, followed by full cleaning and passivation
- After shop hydrotesting wherein the components are drained, air blown dry to a relative humidity end point of 35% or a vapor phase corrosion inhibitor (VPCI) applied, and immediately followed by sealing the equipment for delivery
- What parts, ancillary equipment, and consumables may be used and a preferred list of suppliers/vendors

❒ AUXILIARY BOILER

Emissions from an auxiliary boiler typically need to be accounted for in the application for the air permits. Therefore, the decision to use an auxiliary boiler as a permanent part of the overall plant design, or to facilitate initial plant start-up, should be made early in the project development phase.

If the system design does not include a permanent auxiliary boiler, rental of a portable low-pressure boiler should be considered at least for initial plant start-up since it can dramatically decrease start-up and commissioning time. Rental auxiliary boilers are available from many suppliers for this purpose. This boiler typically will be oil or gas fired and will be self-sufficient, requiring no instrument air. Fuel and a supply of filtered, softened water must be supplied. The auxiliary boiler will be used to steam blow the low-pressure steam lines to all end users (i.e. tank heaters, regeneration heaters, spargers, deaerators, etc.). The temporary steam supply will suffice until the main HRSGs are commissioned.

Installation of a permanent auxiliary boiler is recommended if the plant is designed to cycle frequently. The auxiliary boiler will allow the plant to hold vacuum overnight, speeding up the return to base load conditions in the morning and avoiding destabilization of system chemistry resulting from air ingress. In colder climates an auxiliary boiler is necessary to ensure that equipment does not freeze during an unscheduled outage.

❒ ZERO LIQUID DISCHARGE SYSTEMS

Designing a facility for Zero Liquid Discharge (ZLD) has become a common requirement for power plant developers. ZLD allows developers to site a plant in a remote area where no viable discharge option exists (such as a river, municipal treatment facility, etc.), but a source of water is available. It also enables plants to be built in areas where obtaining a wastewater discharge permit is difficult, or where the discharge permit would severely limit the usage of some chemical compounds, compromising reasonable operating efficiency of the unit. Designing the plant to accommodate ZLD has, in some cases, helped promote good community relations while addressing local environmental concerns.

The key to designing a successful power plant with ZLD is to emphasize that the Total Plant Water Management Plan provides the only pathway to a successful design. From the beginning of the development stage it is essential that all plant water system design efforts, including cooling water, wastewater, and HRSG water be fully integrated into the ZLD concept. Plant designers need to coordinate with water treatment equipment manufacturers and water treatment chemical suppliers as early as possible to develop the water management plan. Special attention should be paid to the types and quantities of chemicals added to the plant water systems, especially products containing dispersants, to ensure that these products do not hinder or limit operation of ZLD system equipment.

When designing a plant for ZLD the problem of wastewater disposal in the event of ZLD equipment problem should be addressed. Typically it is impractical to design 100 percent redundancy into the ZLD system. Therefore, plant developers must provide an alternate means of dealing with wastewater when equipment maintenance or replacement is required. Unless an emergency wastewater discharge location is available, it is recommended that a storage tank or wastewater pond be included in the plant design and sized to hold a minimum of 48 hours of wastewater to deal with these situations. The ZLD system should then be sized to empty this emergency reservoir in a reasonable period of time without hindering normal full-load operation.

❒ MISCELLANEOUS DESIGN DETAIL/OTHER EQUIPMENT CONSIDERATIONS

- Specification of tie-in connections for all temporary flush and fill lines, drain lines, steam blow lines and mufflers, chemical feed lines, sample lines, and the sizing of auxiliary steam preheat lines for cleaning, flushing, and steam blows should be considered during detailed design. All drains, vents, chemical cleaning, and steam blow connections should be sized and located to facilitate pre-commissioning. Tie-in connections must have double block-and-bleed valves that are easily accessible and sized to provide acceptable equipment evacuation rates.

  Large connections are needed for evaporator circulation and vents should be taken directly off the steam drums for chemical cleanings. Consideration should also be given, both in terms of sizing and location, to the ability to evacuate all circuits during cyclic operations and system outages. Superheaters and reheaters should have large drains on the ends and the central portion of the headers for purging accumulated condensate during an outage.

- Makeup water must be available for hydrotesting, cleaning, flushing, steam blows, and performance testing. Makeup water quality specifications are provided in Table 5a, 5b and 6. The quality of first fill water should be the same as that required for the HRSG during performance testing in order to minimize the likelihood of delays during testing.

- Provisions should be made for the temporary feeding of chemicals to storage tanks, surge tanks, and the condenser hotwell when high-volume water usage rates are scheduled.
- Steam from an auxiliary or rental boiler, together with steam injection lines and mufflers sized to effectively preheat the high-purity water storage tanks to a sustainable 180°F will be required during the steam blow process.
- Methods for equipment storage for use between pre-commissioning operations and during shutdowns can be found in references 1, 19, 28, and 29.
- The following documents must also be developed: Pre-commissioning and Commissioning Procedures, Operational Procedures, Operations Training Manual, and an Operations and Maintenance (O&M) Manual. Operations personnel must be trained using these procedures and manuals. Manufacturing and construction cleanliness specifications[31] addressing shop testing, water quality, cleaning, packaging for shipment, site storage and preservation of equipment should also be developed.
- Other design considerations can be found in Section 4 of the ASME document "Consensus on Operating Practices for Control of Water and Steam Chemistry in Combined Cycle and Cogeneration Power Plants"[17].

TABLE 5a

Water Quality Selection – Solvent Based Chemical Cleaning

	Shop Hydro[1]	Site Hydro[1]	Flushing	Alkaline Degreasing[2]	Chemical Cleaning	Neut./Flush	Passivation	Storage	Steam Blow	Storage
Filtered or potable water[3]			A		A[4] (inhib. HCl or HF)			Follow dry lay-up[5] procedures		Follow dry lay-up[5] procedures
Filtered or potable water[3] with pH adjust	A	A	A							
Demineralized			A	R[4] (TSP)	R[4] (Organic Solvents)					
Demin with pH adjust[6]	R	R	R			R	R (+O_2 scav.)		R (+O_2 scav.)	

R – Recommended
A – Acceptable in some cases
Gray – Unacceptable
HCl – Hydrochloric Acid

HF – Hydrofluoric Acid
TSP – Trisodium phosphate
Organic Solvents – Ammoniated ethyene diamine tetra acetic acid or citric acid

Note: Water should not be left in the equipment longer than 24 hours after each unit operation.
1. Non-drainable components constructed of stainless steel should only be filled or flushed with water that meets normal operating limits.
2. Required to remove barrier coatings if preservatives were used.
3. Filtered/Potable water is specified in Table 6.
4. Demineralized water dosed with an amine or ammonia must be used for backfill of superheaters provided that they can handle the additional weight.
5. Lay-up with oxygenated water is NOT recommended. For lay-up procedures see OEM procedure and/or discussion in this document.
6. pH is adjusted to >10 with ammonia or amine

TABLE 5b

Water Quality Selection – No Solvent Based Chemical Cleaning

	Shop Hydro	Site Hydro[1]	Heated Alkaline Flush	Rinse	Storage	Steam Blow	Storage
Filtered Water (Potable)			NOT RECOMMENDED				
Filtered Water (Potable) with Treatment							
Demineralized							
Demin with pH adjust[2]	R	R		R	Follow dry lay-up[3] procedures	R (+O$_2$ scav.)	Follow dry lay-up[3] procedures
Demin with TSP			R				

1. Non-drainable components constructed of stainless steel should only be filled or flushed with water that meets normal operating limits
2. pH is adjusted to >10 with ammonia or amine
3. Lay-up with oxygenated water is NOT recommended.

TABLE 6. Water Quality Specifications for Pre-commissioning Activities

Constituent	Filtered/Potable Water	Demineralized
Conductivity, μS/cm		<2.0
pH, standard units	6.5 – 8.5	5.5 – 8.5
Sodium as Na, ppm		< 0.10
Chloride as Cl, ppm	<100	< 0.10
Fluoride as F, ppm		< 0.10
Silica as SiO_2, ppm		< 0.05
Suspended Solids, ppm	<1	<1
Turbidity, NTU	<1	< 0.5
Total Hardness as $CaCO_3$, ppm		<0.01
Total Dissolved Solids, ppm	<500	
Caution! Equipment manufacturers often issue much more stringent specifications for demineralized water purity that must be considered and adhered to.		

□ Section 6 □

OFF-SITE FABRICATION AND ASSEMBLY[32]

Shop hydrostatic testing water should be pH adjusted and contain a liquid-based VPCI. After shop hydrostatic testing, components should be drained and air-blown dry. A vapor-phase corrosion inhibitor (VPCI)[19] may be applied prior to capping and sealing for transportation and storage.

Regardless of the cleaning and protection procedures used during off-site fabrication and assembly, visual and videoscopic inspections should be performed in components that may have collected moisture (drums, headers, tubes) during the fabrication, storage, construction, or commissioning processes. Section 10 contains information on the inspection and cleaning procedures for HRSG units.

❐ Section 7 ❐
ON-SITE CONSTRUCTION[30, 31, 32]

❐ DELIVERY AND STORAGE

Following delivery of the off-site fabricated components, the OEM's turnover checklist should be reviewed with the field erector. A suggested minimum list of items to include in the turnover checklist is provided below in the Site Receivables Check List.

Site Receivables Check List and Procedure:

- Verify and ensure that all materials and equipment as received compare with the shipping documents and the bill of materials on the purchase order.
- Inspect all materials received for cracks, dents and missing or broken parts.
- Inspect all materials received for damaged or missing closures (covers and caps). Where closures are missing or damaged, the components should be inspected for evidence of dirt and/or water and any resulting damage. After the damage inspection is complete, any missing or damaged coverings should be replaced or repaired prior to storage.
- Materials shipped by water should be inspected for salt accumulation and corrosion.
- Materials, particularly tube bundles, should be inspected for evidence of shop hydro-test water remaining in the assemblies. It is important to check for water in these assemblies to mitigate potential corrosion or damage caused by water freezing in the assemblies.
- Materials should be dispersed to specific storage areas based on sensitivity to temperature (liquids, resins, membranes, and chemicals), exposure to rain or other precipitation, and to comply with the manufacturers' site storage guidelines.

Sealed components installed with corrosion coupons should have the primary coupon removed for corrosion rate determination and then be resealed for site storage. At the time of assembly, the secondary corrosion coupon should be removed for corrosion rate determination. The results of the corrosion evaluation should be documented and made available prior to the site hydro-test to allow ample time for determining what type of post-hydro cleaning may be required. The quantity and quality of any fluids drained when components are lifted into position should be documented.

Any deficiencies or damage should be noted, documented and immediately brought to the attention of the manufacturer for correction. In addition, the results of the initial inspection should be provided to both construction and start-up personnel since corrosion or contamination discovered during the equipment receiving process may impact plans for cleaning of the equipment during a subsequent phase of the project.

Maintaining cleanliness of system components during on-site storage and erection may avoid the need for more aggressive chemical cleaning, minimize the length of steam blow activities, reduce volumes of liquid waste, and minimize the length of time required to meet acceptance criteria during plant start-up. The following guidelines are appropriate for maintaining cleanliness during on-site storage:

- Include a contractual agreement for on-site storage/handling, including an "acceptable cleanliness" specification and penalties for nonconformance.
- Ensure that construction and technical start-up teams work together to educate employees on the importance of cleanliness and good housekeeping in minimizing the potential for additional corrosion and fouling.
- Establish a controlled environment warehouse on site.
- Generate, post, and maintain an updated list showing what components require "outdoor storage", "indoor open storage", and "environmentally controlled storage".
- Do not open, unwrap and leave exposed sealed components during on-site storage. Unsealing and unwrapping is to take place only upon placement for immediate erection.
- Make every effort to minimize the infiltration and retention of water, dirt and mud into or onto "outdoor storage" components.
- Establish a routine surveillance program to regularly inspect all components to ensure that protection devices including covers, sheeting and paint remain in place and free from deterioration.
- Keep the outdoor storage area and surroundings devoid of vegetation and covered with inert ground cover, minimum 3/4" x 3/4" crushed/grated stone only with no sand. This will prevent collection of ground water during rain events.
- Adequately support large outdoor storage components, such as HRSG modules and drums, on proper dunnage or blocking to prevent damage and accumulation of water.
- Mow or cut grassed and weeded areas as frequently as required to minimize airborne contamination.
- Spray soil/dirt areas with water or lime solution to suppress airborne dust.

Other items that should be verified during the on-site construction phase are as follows:

- All debris, including welding rods and other small items, must be removed from the condenser before the condenser walls can be hydro-blasted prior to hydro testing.
- The instrument air compressors and dryer sets must be complete. Disconnect all instrument supply lines and air-blow the entire air system using both instrument and plant air. Load the desiccant and test the dryer regeneration system. Reconnect all instrument supply lines and loop-check all valves and instruments. Air-blow the instrument air system again with dry air [-40°F (-40°C) dew point].

- Site electrical power systems must be inspected for readiness and be available to supply power as required. Trained and competent electrical personnel must inspect and certify all electrical systems and connections. All electrical power safety devices must be functional and available
- All sewer systems must be available and ready to accept wastewater, including spent regenerants, flush water, sanitary sewer water, blowdown water, etc.
- Although the initial boiler hydrostatic test will require demineralized water, the following systems may be flushed with lower quality water, such as filtered or potable water, as specified in Table 6.
- All underground piping
- Cooling water lines
- Fuel oil system
- Natural gas lines
- Firewater system
- Storage tanks
- Pre-commissioning and commissioning procedures must be developed. Completion of the required operating procedures, operations manual, and operator training should be performed during the site construction phase. OEM representatives are to be available where and when required to ensure proper commissioning of their equipment, procedure writing, and training of site operations personnel.

☐ CONSTRUCTION/COMMISSIONING

- Mechanical and chemical cleaning of installed system components[31,32,39] should include general debris removal, hydro-lazing, high velocity flush (8 - 15 ft./sec or 2.4-4.5 m/sec), and chemical cleaning.
- Ensure waste disposal pathways are available during system cleaning. If cleaning waste is to be discharged to the site waste treatment facility, this information must be included in the discharge permit. Otherwise, off-site disposal will be required since sending waste cleaning solvents/reagents to the site waste disposal system will not be an option. Flush waters may still be acceptable for discharge via the site system.
- The HRSG chemical cleaning plan should include degreasing, rust and other oxide removal, and passivation. A video inspection may be performed to determine if any additional cleaning is required or if adequate cleaning, passivation, and storage methods were applied to all shop fabricated sections prior to field erection.
- The condenser should be cleaned, with all debris removed and hydro-blasted prior to filling and hydrotesting.
- All flush and hydro tests are to be performed with suitably pure chemically treated water (Tables 5a, 5b and 6). All water should be drained within 24 hours of test completion. Flush water cascading is allowed (from high to low purity) provided that the specified pH is maintained throughout the flush. For example, water consumption can be minimized by flushing the condenser hotwell forward to the feedwater system, the feedwater system forward to the steam generator and, finally,

the steam generator forward to the condenser cooling water circuit. The flush water from each circuit must be filtered (preferably with 1 micron rated filters) to achieve Table 6 quality water prior to reuse in the next circuit.

- The results of all inspections, photographs, corrosion coupons, water quality sample analyses, virgin and spent solvents, and performance testing from the following activities should be documented and retained.
- Off-site fabrication and assembly
- On-site storage and construction
- Water treatment plant start-up
- HRSG monitoring and control systems, sampling system, and on-line analyzers
- All piping and water storage tanks associated with the plant (e.g., cooling water lines and sumps, fuel oil system, natural gas and compressed air lines and fire water system)
- Hydro tests and high-velocity flushes
- Preoperational cleaning of HRSG
- Steam blows
- Lay-up periods between pre-commissioning processes
- Steam turbine and steam purity tests
- Baseline performance test run

❐ SCHEDULING CONSIDERATIONS

- OEM representatives should be available on-site when required. These representatives can ensure proper commissioning of their equipment and can be called upon for training purposes and procedure development. The selected water treatment and services provider should be on site during all pre-commissioning and commissioning activities. The bid specification should clearly define the contractual commitment of the OEM's representatives, including the number of days on site to support pre-commissioning, to meet performance criteria with the EPC technical team, and for "sign off" of inspections.
- If there are delays during commissioning, or delays between pre-commissioning phases, the HRSG should be stored using suitable corrosion inhibitors and a nitrogen cap.
- Chemical cleaning should be performed immediately prior to start-up of the HRSG. If there is a delay between chemical cleaning and start-up, the HRSG must be laid up in accordance with the procedures discussed in Section 14. An inspection should be performed after chemical cleaning, but prior to start-up of the HRSG. Oxide removal and repassivation may be required. Makeup water treatment equipment should be commissioned and made available to supply high-purity water several weeks before cleaning and flushing activities are scheduled to begin. Consideration should be given to utilization of rental high-purity water treatment trailers to handle the high-volume flushes and steam blows.

❐ Section 8 ❐

MAKEUP WATER TREATMENT PLANT START-UP

The pretreatment system should be placed in service prior to the demineralization system. Systems with clarifiers or precipitation softeners can require two or more weeks of operation before the effluent water quality is suitable for supply to the demineralization system. Once the pretreatment system is functional and providing satisfactory quality water, the on-site demineralization system can be commissioned. If the pretreatment system is performing poorly and demineralized water is needed, mobile demineralization equipment should be used to avoid fouling resins or membranes in the new demineralization equipment.

The demineralization system should not be commissioned until all on-line monitoring instrumentation is installed, calibrated, and operational. During certain phases of the commissioning process, mobile demineralization equipment can be used to supplement the plant demineralization system. Preoperational flushing and cleaning operations can use large quantities of demineralized water which may not have been considered when the makeup treatment and storage systems were sized. Rental high-purity water treatment trailers may be required to handle the high-volume flushes and steam blows. If the capacity of the rental units is insufficient to handle demand, additional mobile stainless steel storage tanks can be rented. High-purity water (delivered in stainless steel tankers) may be required to fill storage tanks.

The following tasks should be performed before the makeup water treatment plant is commissioned:

- Inspect the entire system prior to handover and develop a complete action list for all system deficiencies and incomplete items. Ensure that all relevant disciplines are involved in this inspection such as the maintenance personnel, operations personnel, engineering personnel, management, etc. Ensure that system deficiencies are corrected before completing the balance of commissioning activities and system turn-over.
- Pre-commissioning and flushing of the make-up water treatment plant equipment should be initiated beginning at the raw water inlet and continuing through all water treatment equipment. (e.g. pre-settling pond, clarifier, filters, demineralization system, and all treated water storage tanks).
- The installation of a site potable water system, including storage tanks, must be completed and the components flushed to remove debris. Commissioning of the potable water storage tank includes disinfection of the tank. The water in the potable water storage tank must be tested to ensure it is disinfected and safe for consumption and potable use. Commission the potable water system and flush all end user points with copious amounts of water. Flush until clean and the effluent contains low turbidity and iron. Sample the potable water at the end of the piping runs and have a Health Unit technician or equivalent, analyze the water. This analysis is performed to verify compliance with the applicable health

department specifications and to ensure that all safety requirements are met for on-site use and human consumption.

- Flush all safety showers and eye wash stations until clean. Ensure that all safety shower-heating systems are commissioned and functional. Test safety shower alarms.
- Take inventory of chemical storage tanks. Verify that the tanks are filled before testing associated chemical feed systems in preparation for system start-up. Chemical storage tanks must be flushed as required prior to receipt of chemicals.
- The control system (DCS or PLC) must be installed and tested. Operator training for the control system must be complete and the operators must be trained for operating and troubleshooting the system as required. Loop-check all system valves and instrumentation.
- Ensure all vessel internals are installed correctly and inspected for proper installation of the internals and correct nozzle orientation. Inspect strainers and/or media traps.
- The OEM representative if present should inspect all vessels and internals prior to loading ion exchange resin, filter media, or membranes.
- The demineralizer regeneration system should be checked for leaks and the demineralizer should be regenerated sufficient times to demonstrate that a fully automatic regeneration can be performed satisfactorily as required.
- The demineralizer neutralization system should be confirmed as complete and ready to receive spent regenerants. The wastewater system must be complete and ready to receive wastewater.
- Flush the regenerant and service headers and all regeneration piping. Load the resin using the OEM's approved method and prepare to backwash the resin to remove resin fines off the bed. Take a small sample of resin from each vessel for future inspection or testing. Ensure the correct volume of resin is loaded and verify that the final level of the resin in the vessel is acceptable per OEM's instructions. It is recommended that the vessels be filled with the appropriate water and the resins soaked overnight to rid the bed of entrained air before regenerating. Backwash at a flowrate high enough to fully expand the bed but not enough to backwash whole resin beads out of the top of the vessel.
- Take inventory of the demineralized water storage tank. Have sufficient demineralized water available for resin regenerations and other end user applications.
- Operate the demineralizer trains to acceptable service endpoints and regenerate as specified. Pay particular attention to backwash rates vs. water temperature, regenerant dosage concentrations, step times, and flow rates.
- A resin charge can be irreparably damaged by improper regeneration and require immediate replacement. Therefore keep a log of all regeneration steps and procedures.
- Verify, by means of any vessel sight windows, that the resin volume retains the specified resin levels shown in the OEM drawings and design documents after each regeneration.

❐ Section 9 ❐

INITIAL HYDROSTATIC TEST (HYDRO) AND HIGH-VELOCITY FLUSH (HVF)

The high-velocity flush is an important cleaning step to:

- Avoid equipment erosion and plugging.
- Remove contamination that leads to localized corrosion.
- Enhance the chemical clean and passivation effectiveness.
- Minimize under-deposit corrosion and system pressure drops.

All chemical injection and sampling points, both temporary and operational, should be available prior to the first fill. The chemical feed systems should also be operational and set up to handle the high volume fill and steam blow requirements. If the feed systems' throughputs are insufficient to handle the required volumes, then temporary pumps and/or temporary totes of concentrated chemicals must be installed. If an auxiliary or a rental boiler is scheduled for use during HRSG cleaning and steam blows, it should be operational and on-line at this time to preheat the fill water and spike steam to the HRSG steam drums. It is likely that leaks will be found during the Hydro. If leaks are found, this will necessitate isolation, drainage, and post-Hydro repair.

Prior to the Hydro, the following items should be also functional:

- Planned flush and chemical cleaning paths, including system isolation valves or pipe blanks.
- All permanent and temporary connections, piping and correctly sized valves to permit isolation of the flush and chemical cleaning pathways, chemical injection, and sampling.
- Condensate and temporary pumps.
- A temporary tank fitted with full-flow filtration equipment to collect and filter the flush waters prior to return to the condenser hotwell.
- Sufficient hydro and flush water inventory and quality (see Tables 5a, 5b and 6).

The initial Hydro and HVF of drainable sections of pipework and installed plant equipment should be performed with demineralized water with pH adjusted to ≥ 10.0 as outlined in Table 5a and 5b. It is preferred that the demineralized water quality be at least equivalent to the design specification quality for combined cycle operation. In order to achieve this, the water treatment plant must be functional or a rented mobile water treatment system must be available. Lay-up chemicals should be added to the first fill water to ensure corrosion does not occur. If deoxygenated water is not available, then either an excess of a suitable oxygen scavenger/metal passivator will be required or a suitable silica-free VPCI should be used in its place.

Segments that are to be flushed include low-pressure, intermediate-pressure and high-pressure sections including economizers (forward-flushed), superheaters (backflushed), feedwater system, evaporators, drums, kettle boilers, reheaters, live steam system up to the turbine emergency stop valve, cold reheat system, attemperation water supply lines, and drain piping terminating at the condenser and condenser hotwell. The hotwell, feedwater train, kettle boiler, HRSG and piping up to the emergency stop valve of the steam turbine should be high-velocity flushed, in that order. For the HRSG, the superheaters/reheaters should be back-flushed and the economizers forward-flushed into the respective drums/evaporators. These flushes must be planned along with temporary piping and connections, drains, filters, and critical valve removal or bypass.

Each flush path should be isolated, hydrotested and the individual lines flushed at velocities of 8-15 ft/sec (2.4-4.5 m/sec). The minimum acceptable bottom-to-top flush velocity is 3.25 ft/sec (1 m/sec), while the top-to-bottom flush minimum acceptable velocity is 1.65 ft/sec (0.5 m/sec). Pumps must be sized to achieve these velocities throughout the flow path circuit. Temporary piping and additional forward-flushing pumps may be required. The flush should be sustained at these velocities until the suspended solids are less than 10 ppm using a 0.45-micron filter. This may be estimated by passing 100 mL of flush solution through a 0.45 micron filter, comparing to the standard charts, and multiplying the result by 10.

Hydro and HVF waters should not be retained in the pipework and/or components longer than 24 hours after the flush has been completed. Provisions for the collection, treatment, and discharge of leakage and drainage of these waters will be required to comply with local discharge regulations. Drained water can be reused if it has been filtered and chemically retreated to meet the original specification. Pre-commissioning flush water should not be discharged into closed cooling water systems or cooling tower basins since such practices have caused problems associated with:

- Deposit formation
- Volatile organic carbon, oil and grease accumulations
- Galvanic corrosion
- Biofouling
- Microbiologically influenced corrosion

After all flush water has been drained and all residual debris removed, visual and videoscope inspections of drums, headers, drain pots, dead legs etc. should be performed to assess the need to manually remove any remaining debris or whether degreasing or chemical cleaning is required to remove iron oxide deposits in units operating below 1000 psig (6.9 MPa). Chemical cleaning is recommended for units operating at or above 1000 psig (6.9 MPa).

⫐ Section 10 ⫐

HEAT RECOVERY STEAM GENERATOR CLEANING

⫐ INSPECTIONS

Equipment inspections often require the HRSG sections to be drained and open. Performing inspections throughout the commissioning process is advised to ensure that debris is removed and deficiencies in waterside surfaces are detected and corrected. Inspections should include the following activities:

- Crawl-through visual inspection of waterside surfaces of all vessels (steam drum, condenser, makeup and condensate tanks, and deaerator, if installed). Cogeneration facilities may have additional vessels that require inspection.
- Videoscopic inspections of headers and a representative number of tubes in each accessible tube bundle. Individual HRSG sections should be inspected immediately before assembly to document the "as received" condition. This inspection will help identify waterlines and corrosion which may be present before assembly.
- Tube sampling. If the videoscopic inspections indicate the presence of significant deposits in the tubes, a tube sample may need to be collected to select an appropriate cleaning solvent.

In many instances, not all of the listed inspections have been performed. However, utilizing all of the above inspection methods is recommended to minimize the probability of encountering unexpected problems during commissioning and later operation of the plant.

⫐ PREOPERATIONAL CLEANING

After Hydro and HVF, the HRSG is "degreased", chemically cleaned (if necessary), re-inspected and prepared for steam blow. Successful cleaning of the HRSG will remove contaminants such as weld spatter, grinding dust, non-protective iron oxides introduced during fabrication, contaminants that may have been introduced during shipment and storage, and the contaminants that can be expected as a result of normal site construction and installation practices. If not removed, these contaminants can combine with moisture, with or without oxygen, and contribute to localized corrosion before, during and after commissioning.

Successful removal of contaminants using an effective preoperational cleaning procedure can be expected to provide shorter plant commissioning time and increased reliability and availability during routine power operation[33,34,35,36,37,38,39]. Therefore, it is prudent to perform chemical cleaning of the entire water/steam cycle. Other benefits of cleaning and passivation include increased heat transfer efficiency during subsequent operation and decreased pressure drop from frictional losses due to the elimination of deposits on system surfaces.

Chemical cleaning is advised and should be performed just prior to steam blows to minimize corrosion after cleaning. After cleaning, the unit should be stored in wet layup, using a volatile pH agent and oxygen scavenger plus overpressure with spike steam or nitrogen. An alternative storage method would be to use a liquid silica-free VPCI.

Table 7 presents an overview of the types of materials to be removed by preoperational cleaning activities. Due to the difficulty of performing effective inspections, the safest approach is to perform a preoperational cleaning on all HRSG units. If all of the steps involved in fabrication, storage and erection were performed optimally and thorough inspection confirms that all system surfaces are clean and passive, chemical cleaning for iron oxide removal and surface passivation may be bypassed.

Pre-cleaning procedures will differ depending upon the cleanliness of the internal surfaces found during the videoscope inspection. Heavy iron oxide mill scales and weld material may require the use of an aggressive solvent, followed by an organic acid/passivation soak and rinse.

Pre-service cleaning of the HRSG is a time consuming multi-step process that must be clearly identified in the start-up schedule. During the planning stage, it should be assumed that preoperational chemical cleaning of the HRSG will be performed. In addition, because heat is required, the cleaning process should be implemented when the HRSG is essentially ready for operation. The following items should be considered during the cleaning planning process:

- Appropriate provisions should be made for heating of the cleaning solution to required temperatures.
- Pathways and temporary piping should be identified and constructed to mate to the already specified and fabricated connections.

The selected cleaning agent and specified water quality inventory should be determined and assured. Inspections will have determined whether an aggressive acid cleaning is required, or whether a hot alkaline/detergent wash (to remove oils, grease, silicone and protective coatings) followed by an alkaline chelant cleaning is sufficient.

Factors influencing the effectiveness of chemical cleaning include temperature, solvent velocity, contact time, solvent selection, and solvent concentration. Traditionally, chemical cleaning of new systems has required a multi-step process, as follows:

1. Water flushing
2. Alkaline degreasing/boil out
3. Chemical cleaning
4. Passivation
5. Post-cleaning Inspection

❏ WATER FLUSHING

Flushing at low velocity with demineralized water is performed to assist the removal of readily soluble contaminants and small size particulate material from the HRSG, and verify that the cleaning system is leak-tight. Typical flush velocities

Table 7. Overview of Preoperational Cleaning Methods for Waterside Surfaces

Material	Manual Removal	Flushing	Alkaline Detergent Cleaning	Iron Oxide Removal and Passivation Stages
Large, inert debris (e.g., tools, cans, lids, plastic, gaskets, welding rods, nuts, bolts, etc.)	C	P	N	N
Large, light, partly degradable debris (e.g., rice paper, other paper, dirty rags)	C	C	P	N
Loose, small, inert particles (dirt, sand, grit, grinding debris, wire pieces, welding slag, blasting materials, siliceous and otherwise)	P	C	P	N
Organic films (cutting or lubricating oils and preservatives)	(a)	N	C (a)	N
Semi-adherent deposit or debris (dirt, loosely adherent mill scale, light and porous tubercles, loosely adherent slag particles)	N	P	P	C (b)
Adherent deposit (mill scale, non-protective scales from welding and heat treatment, tubercles, and general corrosion product layers)	N	N	N	C

C - Complete removal possible
P - Partial removal
N - No effect

(a) Grease may require manual removal with alkaline detergent
(b) By sloughing of underlying oxide

are at least 1.6 ft./sec (0.5 m/s) for a top-to-bottom flush and at least 3.3 ft./sec (1 m/s) for a bottom-to-top flush[30]. The end point for the water flush has two components, conductivity and suspended solids. Typical end points are:

- Conductivity ≤ 2 μS/cm above the conductivity of the influent water
- Suspending Solids <10 ppm

❑ ALKALINE DEGREASING/BOIL OUT

In most cases, alkaline degreasing is performed at elevated temperature (e.g., >180°F or >82°C) for 12 – 24 hours. Higher temperatures are preferred where possible. The preferred degreasing solution contains a mixture of tri- and di-sodium phosphates and a surfactant (wetting agent). Some HRSG manufacturers may recommend, or require, the use of a high-velocity flush in place of the low-velocity flush followed by alkaline degreasing for the associated piping. Sodium hydroxide with a surfactant has also been used for alkaline degreasing of many HRSG units.

Alkaline boil out is performed to remove oil, grease, preservatives, and other organic materials from the wetted surfaces of the HRSG. The same cleaning chemicals that are used for degreasing the associated piping are also used for performing the alkaline boil out. The required length of the boil out will vary from 8 to 48 hours depending on the types of HRSG preservatives utilized. Follow the OEM instructions for boil out duration and pressure.

❑ CHEMICAL CLEANING

Some manufacturers require that the HRSG is chemically cleaned before being placed in service. In other cases the need for chemical cleaning is determined by inspection. A good visual inspection necessitates the use of high-definition videoscope (fiber optic and/or micro-camera equipment). Accurate interpretation of the video data is highly dependent upon the experience and skill of the analyst. Removal of sections of tubing for laboratory examination may be required if unambiguous interpretation of the video data proves impossible. Therefore, conservatism dictates that preoperational cleaning of the HRSG is the recommended practice. It is recommended not to use the condenser hotwell as a reservoir for the chemical cleaning solution because of the potential for turbine contamination.

Many generic and proprietary formulations for HRSG chemical cleaning solvents are available. A comprehensive discussion of these formulations, functionalities, and application methods can be found in the literature[18]. Most of these formulations can be grouped into one of three types:

- Mineral acid
- Organic acid
- Chelant

When selecting the process that will be used for preoperational cleaning, it is important to remember that the deposits present in the new steam generator were not formed at operating temperature and pressure. This means that the elevated temperature solvent conditions required to remove reducible metals and/or their oxides and hydrothermal minerals are not needed. Additional details on the various aspects of preoperational cleanings are available in the literature.

❏ PASSIVATION

The method of achieving passivation after chemically cleaning the iron surfaces is highly dependent upon the solvent used in the cleaning process. All passivation procedures are enhanced by maintaining good fluid circulation. It should be noted that the iron oxide film established during the implementation of these post-chemical cleaning passivation processes is largely cosmetic. The magnetite that develops provides protection against flash rusting, however, the layer of magnetite formed is not sufficiently dense, or well enough structured, to provide protection against the migration of potentially corrosive species through the oxide film. Strong magnetite films formed at operating temperatures are generally much better than those that can be formed in cleaning passivation stages. A high temperature (e.g., 300-400°F or 149-204°C) boil out with an oxygen scavenger and an alkali to keep pH values of 10-11 has been used to form a more adherent passive layer. Should poor passivation be noted during the post-cleaning inspection, an additional passivation step consisting of a high-temperature boil out with pH-adjusted demineralized water and containing a high concentration of an oxygen scavenger should be considered. An alternative, lower-temperature process with pH-buffered demineralized water and an elevated concentration of sodium nitrite may be used, but would require a much longer contact time.

❏ POST-CLEANING INSPECTION

A comprehensive visual inspection should be performed after the chemical cleaning process has been completed. This inspection is used to verify that no unwanted deposits and/or debris remain in the HRSG and that the cleaned surfaces are coated evenly with a layer of adherent magnetite. All inspection observations, photographs, water and deposit analyses should be documented and the results reported. If the surfaces appear to be coated with loose black magnetite, it is recommended that a sample of this material be removed and analyzed. Depending upon the planned time between the end of cleaning and first fire/steam blows, the HRSG may be steam-dried using an auxiliary boiler or preserved utilizing a wet or dry lay-up technique. The estimated time delay and anticipated weather conditions (freeze potential) prior to firing for steam blows should be considered to determine method of storage. Lay-up should be maintained until steam blow. In all cases, nitrogen capping (3 to 5 psig or 0.02 to 0.035 MPa) is required.

- If wet lay-up is selected the HRSG should be prefilled using a nitrogen purge prior to filling with treated water. The acceptable pH (>10.0) and scavenger/passivator concentration must be maintained throughout wet lay-up. During steam blows the lay-up chemicals are either consumed or removed with the steam. This avoids the need to drain and dispose of the wet lay-up solution as part of the chemical cleaning waste inventory. Other lay-up methods may employ the use of vapor phase corrosion inhibitors in a wet method application.

- If dry lay-up using desiccant is selected, a recorded number of desiccant containers should be placed in the water side of the HRSG. The use of silica-free liquid VPCI fogged into the vessel for dry lay-up may also be evaluated.

❏ WASTE DISPOSAL

The amount and chemical composition of waste generated during chemical cleaning is highly process dependent. For a preoperational steam generator chemical cleaning, the metals content of the waste is typically less of a problem than with post-operational chemical cleaning applications.

Before beginning the chemical cleaning, it is important to ensure that there is sufficient tank storage space available to hold all of the spent process fluids and used rinse water. It is also necessary that the local wastewater discharge limits are reviewed. On-site treatment and disposal of the spent solvents (and some of the rinses) is often not an option and the wastes must be hauled offsite by a licensed cleaning waste handling contractor.

A waste handling, treatment and disposal plan must be developed and approved by your environmental and safety coordinators before proceeding with the cleaning process. For facilities with on-site treatment, the appropriate waivers, environmental permits, and protocols must be in place prior to the start of the cleaning. Discharge of spent cleaning solvents to wet surface air condensers that contain galvanized coatings is discouraged because the protective galvanized coating may be lost. Detailed waste handling, treatment and disposal options are beyond the scope of this document.

❏ HRSG CHEMISTRY MONITORING AND CONTROL SYSTEMS

Following completion of preoperational cleaning activities, chemistry monitoring equipment should be fully operational in preparation for steam blows, performance testing, and plant start-up. The on-line analytical instrumentation should remain isolated until steam blows have been completed.

The testing laboratory should be stocked and functional. All sample lines, including temporary lines, should be labeled and operable with adequate flow and cooling available[23] and the delay time response calculated. Sample lines for iron and copper corrosion product monitoring should be constructed of stainless steel and have diameters designed to obtain fluid velocities of 5-6 ft/sec (1.5-1.8 m/s) for accurate results. Prior to start-up of sampling equipment, the individual sample valves should be opened one at a time, purged, and tested to verify correct labeling and routing.

The chemical treatment programs[15] should be determined and water chemistry control limits[17] should be in place. Chemical treatment programs are fully delineated in the ASME "Consensus on Operating Practices for Control of Water and Steam Chemistry in Combined Cycles and Cogeneration Power Plants".

All automatic control loops, support systems, sensors, interlocks, pumps etc. should be commissioned and DCS verification performed. Any automatic lockouts for valves during trips (e.g. automatic lockout of blowdown valves requiring manual reset, control valves and level sensor set points corrected to operational temperature/specific volumes etc.) must be identified and their effect on the steam blow and performance testing process reassessed. For units with deaerators, the deaerator must be made operational at this time and a functional auxiliary boiler needs to be available as a source of start-up steam. A action list should be established in order to avoid delays during steam blow, steam turbine start and baseline testing.

Additional items worthy of consideration include:

- Chemical feed systems must be operational and in automatic mode.
- The water treatment plant instrumentation and control system (DCS or PLC) should be complete, commissioned, and tested. Operator training for the system must have been completed.
- Information on sampling locations, sample conditioning and analysis/analyzers for the monitoring of system cycle chemistry can be found in the ASME "Consensus on Operating Practices for the Sampling and Monitoring of Feedwater and Boiler Water Chemistry in Modern Industrial Boilers"[24].

❑ Section 11 ❑

STEAM BLOW

Air blows do not thermally cycle the steam circuit piping and, therefore, are not a recommended substitute for steam blows.

The production of a good protective passivation layer on the wetted surfaces within the steam generation system requires elevated temperatures and contact time. Because steam blows are quite lengthy, demineralized water containing a volatile oxygen scavenger/metal passivator and a volatile alkalizing agent should be utilized during this part of the pre-commissioning process. This combination facilitates removal of particulates from the after-boiler piping.

The steam blows are continued until each of the target surfaces (typically three) reveal no raised marks, less than 5 unraised pits, and impingement marks less than 0.5 mm in diameter on two consecutive runs lasting a minimum of 15 minutes each. The longer it takes to achieve acceptance, the more costly the process. The duration of the steam blow will vary depending on the number of paths chosen for each drum, the cleanliness of the system, and the pressure used for the steam blow. Extended steam blows may be required if off-site fabrication, component storage, and/or site pre-cleaning implementation were inadequate.

In an attempt to minimize the time required for steam blows, the ability to dislodge particles can be increased by operating at higher design cleaning force ratios (CFR). The CFR is a measure of the ratio of the drag force produce the blowing process to the drag force produced under maximum normal operating conditions or maximum continuous rating (MCR). The CFR is typically calculated as follows:

$$CFR = \frac{G^2_{blow} \times V_{blow}}{G^2_{mcr} \times V_{mcr}}$$

Where G = steam mass flow for the blow or MCR

V = steam specific volume for the blow or MCR

If the calculated CFR is greater than one, any particulate matter not removed by the steam blow should not be dislodged during normal operation. Typically, design CFR values between 1.1 and 1.7 are used. It is the pressure, and hence the specific steam volume, and the steam mass flow during the blow that are adjusted to provide CFR values greater than one for a fixed steam line size.

The flow paths are typically planned during the design phase and provisions made for the required pressure, line diameter, and terminal discharge location. Reheater, wet surface air cooled condenser, and air-cooled condenser flow paths are the hardest to clean and require longer blows. Three or four thermal cycles of 4-hour blows are more effective for oxide removal than continuous steam blowing. The cool-down between blows is limited to a differential saturation temperature of 75°F (24°C) between the initial and final drum pressures. A reducing and/or passivating agent should be used during the steam blow process regardless of the chemistry control program that will be used during normal operation. The recommended water quality and treatment for steam blows are provided in Tables 5a,5b and 6.

Other important planning considerations for steam blows include:

- Provision for temporary high-volume feeding of chemicals to storage, surge tanks, and condenser hot well.
- Steam injection mufflers and supply lines sized to preheat the high-purity water storage tanks to a sustainable 180°F (82°C) during the steam blow process.

Prior to commencing the steam blows, removal of the internals of designated valves and verification of an adequate supply of demineralized water should be completed. The deaerator and demineralized water storage tank preheater (with a temporarily oversized steam muffler installed to cope with the mass flow of makeup to be used during the steam blows) must be on line. Chemical priming and high-volume chemical feed pump settings should be adjusted to match the anticipated flow rate of demineralized water for makeup. Treatment chemicals should be temporarily added as far back in the system as possible, such as to the condensate storage tank, if practical. Typically, a volatile alkalizing agent and oxygen scavenger/metal passivator are added. During steam blows, grab samples from the feedwater and HRSG should be routinely collected and analyzed to track the elimination of metallic particles via blowdown. It is important that the dissolved oxygen be lower than saturation upon entering each thermal squeeze point prior to the vented low-pressure drum. An example of the steam blow process for a 1x1-configured 250 MW plant with a CFR of 1.4 to 1.6 is provided in Table 8.

Table 8. Steam Blow Process Summary[1]

Pressure		Saturation Temperature	
PSIG	MPa	°F	°C
45	0.31	292	145
90	0.62	331	166
110	0.76	344	173
125	0.86	353	178
185	1.28	382	194

[1]Conditions of steam blow:
- 8 flow paths defined
- Expected flow path duration of blows ranged from 15 to 20 hours
- Steam flow of 50,000 to 320,000 lbs/hr (22,680 to 145,150 kg/hr), depending upon flow path
- Sequential circuit blows, low-pressure first, then intermediate-pressure and finally high-pressure

Upon completion of the steam blows, the HRSG should be hot-drained with a residual 25 psig (0.17 MPa) pressure present, inspected and refitted with typical operational piping, connections, sensors, controllers, valves etc. It should then be laid up as soon as possible using an appropriate method [19,29]. The appropriate chemicals at recommended dosage and spike steam from the auxiliary boiler can be employed if the lay-up is of short duration. If the lay-up time is extended because of site process issues, wet lay-up[19,29] or a liquid VPCI should be used. An adequate supply of steam from an auxiliary boiler must be available if the potential for freezing is a concern. A dehumidified air dry-out followed by the use of desiccants and the installation of a nitrogen overpressure blanket, or the application of a powdered VPCI may be utilized; however, these lay-up techniques are used relatively infrequently because of their poor efficiency.

☐ Section 12 ☐

TURBINE

Turbine pre-cleaning is performed during assembly. However, if residual greases and oils are present it may be advisable to implement a pressure wash prior to assembly

After completion of the steam blows, the temporary piping, valves and controllers are removed and refitted or replaced with those required for normal operation. The steam turbines also undergo readiness for testing by the manufacturer. At this time the calculations for steam purity, feedwater and pretreatment control limits should be revisited[18,41]. This review will allow early identification and correction of steam purity and water quality issues prior to performance testing and help minimize the potential for prolonged delays.

By this time, all plant systems should be complete and available. If permanent in-line condensate polishers are not provided, the installation and use of temporary precoat filters with deep bed, mixed resin demineralizers installed in a side stream around the hotwell will reduce system cleanup time prior to baseline and performance testing. This practice is encouraged, particularly for plants with air-cooled condensers, because steam blow cleanup performance is highly variable.

All the sampling lines and in-line instrumentation must be working and calibrated with the delay time response verified for the appropriate sampling flow rates[23]. The factory acceptance test must be completed and signed off for all control panels. The chemical pumps must be calibrated for the calculated dosage delivery rate. All predetermined chemistry[18,22,24] and chemical treatment[15] control ranges for achieving the desired water and steam purity levels[18,24] should then be verified while initiating the system cycle cleanup prior to baseline testing. The previously calculated operating chemistry ranges for the selected chemical treatment regime should also be available to assist in the interpretation of any out-of-specification steam and/or water purity results obtained during performance testing. This action will minimize long or unnecessary cleanup times prior to initiating the steam turbine portion of the test.

The use of an inorganic AVT chemistry control program, such as hydrazine plus ammonia, may simplify initial plant start-up regardless of HRSG configuration (drum units or once-through units) even if the unit is to be operated on another chemistry regime during normal operation. However, during plant commissioning there is an increased likelihood for condenser tube failure and contaminant ingress. Condensate chemistry, therefore, needs to be closely monitored during this time and appropriate corrective actions quickly employed if a leak is detected, up to and including stopping the commissioning activities. If contamination is severe and timely corrective action is not taken, extensive corrosion damage to system components may be initiated. The use of temporary or rental condensate polishers can help to mitigate some of these concerns.

Conversion of units from the commissioning AVT program to an alternative operational chemistry regime should not be performed until after performance testing is complete. Conversion from an inorganic chemistry regime to an organic chemistry regime after performance testing has been completed may also be an option[16]. However, if organic chemistry is to be used, the feed of those chemicals should be carefully controlled to maintain water and steam within cation conductivity specifications. A typical starting point would be equilibrium feedwater concentrations of ≤500 ppb active organic alkalizing amine plus ≤50 ppb active organic oxygen scavenger/metal passivator. It should be noted that the OEM or EPC Contractor may have different views on the chemistry to be used than the owner/operator. Therefore, a transition in the operating chemistry regime may need to be delayed until after the Owner signs off on the plant or the OEM's guarantees and warranties expire.

The chemistry control guidelines developed by various industry groups address steady-state operations and unit restart after the commissioning and initial start-up period. These guidelines typically include action levels which outline acceptable chemistry deviations based on the hours of operation outside the recommended normal cycle chemistry limits. These limits contain safety margins to minimize corrosion and deposition within the system. However, during initial start-up, the control of cycle chemistry is difficult and achieving the normal (N) control limits can be a lengthy process as the unit cycles on and off routinely during commissioning activities.

It has been demonstrated during numerous start-ups, that initial operation of the turbine at full load will remove impurities hidden in the steam turbine and the HRSG feedwater systems. During full load operation these contaminants dilute out and uniformly mix within the feedwater and condensate systems. This cleanup approach is more effective than blowdown holds, since the continuous blowdown systems are only sized for 2-4 % MCR flow. Therefore, the quicker that turbine full load can be reached, the sooner the target feedwater chemistry and steam purity values may be achieved[42].

It is in the best interest of the owner/contractor to negotiate relaxation of chemistry limits during equipment contract negotiations in order to avoid long holds and costly delays, particularly when steam is first admitted to the turbine. A recommended approach is to specify that contaminant concentrations in the first steam entering the turbine not exceed four times the normal operating limits; these levels should be roughly equal to the steam turbine OEM's Action Level 3 category. The contamination constituent limits should be trending downward before steam is admitted to the turbine and the recommended normal limits are to be met within the negotiated time frame.

For example, Table 9 represents a typical set of steam purity guidelines for plants with drum boilers using all volatile treatment with reheat and a condensing turbine. These guidelines are provided as an example and owner/operators should consult and use actual supplier guidelines in all cases. The action level criteria and maximum cumulative exposure chart specified by this steam turbine OEM[43] allows up to eight hours of operation with chemistry parameters in Action Level 3; parameters must be trending toward Action Level 2 within one hour. If 4 hours are expended in Action Level 3 to reach Action Level 2, then the time allowed in Action Level 2 to reach Action Level 1 is now 24 hours (i.e., one-half of the 48 hours allowed). Similarly, if half of the time allowed for Action Level 2 is used to achieve Action Level 1, then only half of the allowed time in Action Level 1 can be used to achieve the normal acceptable target limit.

Table 9. Example of Steam Purity Guidelines for Drum Boilers with Reheat Using All Volatile Treatment

Target Parameters	Sample	N	1	2	3
Sodium, μg/L as Na	C	≤2	>2 ≤4	>4 ≤8	>8
Cation Conductivity (or Degassed)[2], mS/cm	C	≤0.2	>0.2 ≤0.4	>0.4 ≤0.8	>0.8
Silica, μg/L as SiO_2	C/T	≤10	>10 ≤20	>20 ≤40	>40
Chloride, μg/L as Cl	T	≤2	>2 ≤4	>4 ≤8	>8
Sulfate, μg/L as SO_4	T	≤2	>2 ≤4	>4 ≤8	>8
Total Organic Carbon, μg/L as C	W	≤100	>100	----	----
Specific Conductivity, μS/cm	T	----	----	----	----

All Conductivity measured at 25°C.

Legend Sample Frequency
C = Continuous
T = Grab, once per shift for troubleshooting
W = Grab, once per week
C/T = Recommend continuous, grab once per shift for troubleshooting minimum

Targets and Cumulative Hours Per Year
N = Normal
1 = Action Level 1; 336 hours
2 = Action Level 2;   48 hours
3 = Action Level 3;   8 hours

Ultimately, these core parameters must be met prior to proceeding to performance testing. The ease with which acceptable chemistry is achieved for feedwater and steam is related to the rigor with which quality control, cleaning and inspection procedures are performed during all phases of fabrication, construction, and installation. Typically, all values in Table 9 can be doubled for a non-reheat condensing turbine. Chemistry limits for a non-reheat, non-condensing turbine are generally further relaxed and will vary from supplier to supplier.

Specific guidelines for the location of sampling points, frequency of testing, sampling procedural requirements, and test methods and interferences are provided in the ASME "Consensus on Operating Practices for Control of Water and Steam Chemistry in Combined Cycles and Cogeneration Power Plants," and references 44 and 45. Continuous monitoring should include the HRSG feedwater pH and cation conductivity, and steam sodium, silica and degassed cation conductivity. Isokinetic steam sampling[40] nozzles should be installed on the saturated steam line for HRSG steam purity verification and on the superheated and/or reheat steam, if present, for steam turbine OEM acceptance.

All steam purity analyzers must be in service and recording while steam is venting to atmosphere via mufflers, or by-passing the turbine to the condenser, in order to determine when the steam is acceptable for turbine operation in accordance with the turbine supplier's chemistry limits and specifications.

Analyses for other species specified by the HRSG and turbine OEMs must be performed using grab samples in accordance with the equipment warranty and contract requirements. Having laboratory equipment and additional analytical instrumentation (such as an ion chromatograph) on site to perform chemistry tests will provide rapid feedback and shorten the commissioning period. Use of an off-site laboratory may require delays of hours to days before chemistry can be confirmed or troubleshooting information can be obtained. An OEM-approved laboratory for analyzing grab samples of high-purity water should have been selected during bid review and contract award, even if this testing is performed on site. Once the chemistry parameters meet specification, repeat grab samples are forwarded to the contract laboratory for verification.

Normal feedwater, HRSG, and steam chemistry limits and specifications should be achieved in no longer than three to seven days[42]. Longer time requirements for this activity are typically associated with poor pre-cleaning of system circuits, in-leakage problems, or the failure to use a properly designed and operated condensate polishing system.

◻ Section 13 ◻

STEAM TURBINE COMMISSIONING

The performance tests start when the HRSG is approved to power up to MCR System chemistry cleanup may be required before admitting steam to the turbine. For example, HRSG operation may be held in a continuous run condition of 70% of base steam load while by-passing the steam turbine until the cycle water and steam chemistry limits meet acceptable operating limits for admission to the steam turbine. The percentage of base load that may be achieved during this run is contingent on the air permits available, venting capacity, or the design capacity of the dump condenser. The lower the percentage power, the longer it will take to clean up the chemistry via blowdown.

Key test parameters are associated with both the chemistry control program selected (see Figure 3) and the turbine OEM's operational limits. A suggested set of key test parameters and limits is given in Tables 10 and 11[42]. These are typical limits representing the requirements of one steam turbine OEM. The actual limits for a project are to be agreed upon with the project turbine OEM. For the parameters noted in Table 11, values should be trending downwards from the limits given in the table before steam is admitted to the turbine.

In order to avoid lengthy delays, the following should be addressed prior to initiating the commercial acceptance test:

- The site-specific chemical treatment program manual is complete and contains the following.
  - A discussion on water-and-steam-related corrosion and deposition issues
  - A description of the operating chemistry regime and the functionality of the chemicals and chemistry to be used to achieve the required cycle chemistry
  - Guidelines and procedures for controlling cycle chemistry and key test parameter control level limits (i.e. N, 1, 2, 3) for the operating pressures and steam turbine employed
  - Recommended chemical analyses (test procedures), instrumentation, and control ranges for key test parameters
  - Contingency guidelines for dealing with cycle chemistry upsets, start-up, and lay-up
- Operator training for cycle chemistry control during operation and troubleshooting has been conducted.
- Verification that all sampling panels and analyzers are working, calibrated and that the time delay response has been calculated.
- Verification of system makeup water storage quality and quantity. If the quality is off-specification, consider recirculation through the high-purity pretreatment equipment until the specification is achieved.
- Start-ups will require higher than normal volumes of high-purity water (up to 20% of the feedwater flow) for desuperheating and to replace

wasted blowdown. Capacity of water treatment systems should be verified and rental equipment available as required to meet demand.

- Verification that all chemical pumps are functional and set to provide optimum output for the baseline load and that "draw-downs" are consistent with the feedwater flow integration signal. If the pumps are automated for feed-forward on flow and an in-line analyzer is used as a feedback correction trim, then the selection of Proportional Integral (PI) or Proportional Integral Derivative (PID) control will be dependent upon the delay time from the chemical feed point to the analyzer's sensor. PID control is generally used when the delay time is three minutes or less. The calculated value must be validated during a steady-state run and factored into the Distributed Control System (DCS).
- A final on-site review must be completed and all piping and instrumentation diagrams (P&ID's) updated to reflect any changes in piping, pumps (size/capacity), equipment, and sampling locations.
- Performance test sign-offs and acceptances by OEMs need to be provided.

During start-up, follow the OEMs' procedures to minimize material stresses and maintain proper drum level to avoid carryover. All drain pots on the superheaters and reheaters and all other drain valves should be opened and drained. These steps will minimize the potential for corrosion, deposition and the time to cleanup because drain pots can contain contaminated liquids that create off-specification water quality or steam purity conditions when released to the system. These contaminants will mix and dilute out as load is increased.

Also during start-up, there may be insufficient vacuum to permit the use of condensate/hot well water for condenser pump seal cooling. Initially the stored high-purity/condensate blend tank water in the condensate storage tank may be used for this purpose. However, the use of this water may cause off-specification

Table 10.   Example of Suggested Initial Start-up Feed Water Chemistry Guidelines for Plants

Parameter	Frequency	Limit
Cation conductivity, μS/cm	Continuous sampling	<0.3
pH, standard units at 25°C	Continuous sampling	9.4 to $\geq$ 9.6*
Dissolved oxygen at economizer inlet, ppb	Continuous sampling	<20
Iron, ppb as Fe	Grab every 4 hours	<5
Copper, ppb as Cu	Grab every 4 hours	<2
*Mixed metallurgy pH should be 8.8 to 9.3		

Table 11. Example of HP and IP Superheated and Reheat Steam Purity Limits Required for Steam Turbine Admission

Parameter	Frequency	Limit
Degassed cation conductivity, μS/cm	Continuous sampling	< 0.45
Sodium, ppb as Na	Continuous sampling	< 12
Silica, ppb as SiO_2	Continuous sampling	< 40
Chloride, ppb as Cl	Grab Daily	< 12
Sulfate, ppb as SO_4	Grab Daily	< 12
TOC, ppb	Grab Weekly	< 200

quality water in the condensate/feedwater circuit. Seal cooling water supply once proper mechanical operation is achieved.

Make sure that all specified dampers, vents and by-passes are open prior to start-up to ensure adequate cooling flows. Vent valves are to be greater than 10%, but no more than 50% open. Steam attemperators are to be engaged in a successive manner (typically at 25% of base load) to achieve cooling within 25°F (14°C) of operational design. When 85% of attemperation valve opening is achieved, the next attemperation valve is employed. Attemperation to within 25°F (14°C) of saturation is forbidden to minimize susceptibility to two-phase FAC, tube-to-tube differential expansion, and damage to the low pressure section of the turbine.

During the performance test, the system should be operated continuously at steady-state to validate design performance. These formal performance tests are very important and have a much broader purpose than to determine whether contractual obligations are being fulfilled. They allow for bench-marking the performance of each sub-unit process under non-upset or uncontaminated conditions. They also provide a reference for future performance reevaluations and can be used to diagnose problems. Future performance evaluations can then be used to identify performance trends, sub-unit process operation changes and the root cause of those changes, and to assist in maintenance planning.

Assuming that there are no warranty concerns uncovered during the commissioning step, migration to the operating chemistry regime of choice (PT, AVT(O), or OT) can be made after steady-state base-load operation has been achieved i.e. if PT or AVT(R) regime has been employed for start-up and commissioning. During this period, the following checks can be performed while other performance tests are being run:

- Verify steam purity once again
- Check condenser deaeration performance and air in-leakage prior to acceptance of the equipment

- Test for phosphate hideout if on PT chemistry control. Determine the maximum phosphate concentration that can be maintained in the drum water until a mass balance loss occurs. Reset the upper control limit to 80% of this value. Do not exceed the maximum recommended phosphate concentration for the treatment program.
- Validate and record all the cycle chemistry test results as well as mass balance chemical dosages. These results comprise the site-specific baseline data set, assuming no air or condenser cooling water in-leakage is present. All leaks must be corrected as soon as possible. These data will be used as benchmarks for future comparative analyses and for troubleshooting.

❒ Section 14 ❒

START-UP, SHUTDOWN, AND LAY-UP

Cyclic operations create stress conditions that can cause environmental and thermal fatigue within the equipment. These operating conditions can also lead to the accumulation of contaminated or high-concentration condensates during cool-downs. Such solutions can promote surface penetration of the materials of construction during periods of flexing and increased stress present during restarts or reheats. As the number of cycles increases, so does the probability of metal failure because fatigue cracks initiate and undergo growth during each successive thermal cycle. These stresses can be minimized by following the OEM site-specific recommendations for cyclic operations.

Lay-up procedures during shutdowns will vary, depending upon the duration of the out-of-service period[19,29]. For short-term lay-up, the pH of the water in the HRSG is ideally elevated to a minimum of 10.0 prior to shut down and the HRSG is isolated to maintain pressure and temperature as long as possible. Steam blanketing from the auxiliary boiler or another HRSG can be used to hold positive overpressure in the drum or when freeze protection is required. If an auxiliary boiler or another source of steam is not available during lay-up or lay-over periods, then a capping nitrogen blanket should be used. The nitrogen is back-filled into the steam drums as the pressure decays below 25 psig (0.17 MPa) with the use of a pressure regulator to ensure that 3 to 5 psig (0.02 to 0.35 MPa) pressure is maintained.

In areas where ambient air conditions fall below freezing, all piping, valves, columns, drains and connections that hold water should be heat-traced and all gas side dampers should be closed. Dry lay-up or lay-over should be employed if the HRSG cannot be maintained hot during the lay-over, or no provisions are available to supply steam to keep the unit hot during lay-up. Some sites have experienced millions of dollars of damage from improper freeze protection. Therefore, if/when appropriate, a site-specific freeze protection program must be developed.

Guidance and procedures for short- and long-term lay-ups can be found in the ASME "Consensus for the Lay-up of Boilers, Turbines, Turbine Condensers, and Auxiliary Equipment"[29].

❒ Section 15 ❒

GLOSSARY OF TERMS

Combined cycle – any combination of a gas turbine, steam generator (or other heat recovery equipment) and steam turbine used to improve cycle efficiency in the power generation process.

Concentration cell – an electrochemical cell in which an electromotive force results from the difference in the concentrations of solutes in the electrolyte. This difference in concentrations defines discrete cathodic and anodic regions.

Condensate polishers – see demineralizer, mixed bed.

Conductivity, specific – electrical conductance of a solution is a general property and is not specific for any particular ion. It is a common method of assessing total salt content of a solution. The standard unit of measurement is microsiemens/ centimeter^2 (μS/cm^2). All references to conductivity assume it is measured at 77°F (25°C).

Conductivity, cation – the amplified electrical conductance of a solution that has had all cations and amines present replaced by hydrogen (H$^+$) ions after passing through a highly regenerated column of hydrogen form strong acid cation exchange resin. All references to conductivity assume it is measured at 77°F (25°C).

Conductivity, degassed cation – cation conductance of a solution that has also been degasified to remove any contribution of carbon dioxide to the electrical conductance of the solution. All references to conductivity assume it is measured at 77°F (25°C).

Damage - condition that causes loss of performance or integrity of equipment.

Deaerator, integral – mounted on and attached to the low pressure steam drum which acts as the feedwater pump storage.

Deaerator, pressure – mechanical device in which a low pressure steam/water mixture of feedwater is stripped of dissolved oxygen to a design concentration of 7.0 ppb. Other noncondensable volatile gases will also be removed.

Deaerator, separate – not mounted on low pressure drum but on a separate feedwater pump storage tank.

Deaerator, vacuum – process equipment used to remove oxygen, air, or other noncondensable volatile gas content from the water passed through it by subjecting the water to a vacuum.

Decarbonator (atmospheric forced draft) – process equipment in water treatment systems used to reduce or remove free carbon dioxide from the water passed through it by contact between the water and a stripping gas, usually air.

Degassifier – see "Deaerator, vacuum or Decarbonator."

Demineralizer, two bed – consists of separate resinous cation and anion exchangers which are usually operated as a primary system.

Demineralizer, mixed bed – unit containing an intimate mixture of cation and anion resins separated and regenerated by special procedures. Usually employed as a polishing step after primary units to produce high purity feedwater makeup. Alternatively, may be used to polish condensate returns to remove ionic impurities and heavy metal corrosion products.

Electrolyte – chemical solution containing ions that migrate in an electric field.

Greensand – naturally occurring Glauconite which is processed to achieve a material that will oxidize and remove iron and manganese when used in conjunction with potassium permanganate as a continuous feed to the influent or as a regeneration medium.

Hydrolazing – water blasting at high pressure over a narrow area to clean a surface. This process is often used to clean piping.

Makeup water, cycle – treated water forming part of the feedwater to the steam cycle power train.

Make-up water, non-cycle – treated water utilized for plant purposes which are not part of the steam cycle – i.e.: cooling tower makeup.

Organic carbon, non-volatile – organic carbon not intentionally added as part of the water treatment chemical regime. Nonvolatile TOC was defined by the ASME "Consensus on Operating Practices for the Control of Feedwater and Boiler Water Chemistry in Modern Industrial Boilers" (CRTD-Vol.34) as an unofficial modification of the TOC test conducted on a sample after atmospheric boiling with the subsequent subtraction of a calculated carbon value equivalent to the carbon content of any nonvolatile organic treatment chemicals.

Organic carbon, volatile – organic carbon present either from volatile organic treatment or from volatile organic contaminants.

Pass(es), reverse osmosis – description of the reverse osmosis system configuration in which the permeate from one pass is the feedwater to a subsequent pass operating in series. Only the permeate from the final pass becomes makeup.

Ramp rate – the rate at which a piece of equipment is subjected to increased service load or the rate at which temperature increase is allowed for the HRSG.

Stage(s) – description of the reverse osmosis system configuration in which the concentrate or reject from one module is the feedwater to a subsequent stage and the permeate from each is combined as a single stream.

Stand-alone plant – a self-contained facility having no external steam host.

Steam host – a plant or facility external to the generating plant that utilizes the steam and returns the steam condensate for recovery and reuse to the generating plant.

Ultrafiltration – membrane system similar to reverse osmosis that is capable of removing particulates, colloidal matter etc. but incapable of removing ionic and other dissolved solids.

Water, demineralized – treated water produced by resin or membrane processes having a specific conductance ≤10 μS/cm at 25°C and containing ≤ 0.01 ppm iron as Fe. When used in non-drainable superheaters or reheaters constructed with austenitic stainless steels this specification shall be ≤ 1.0 μS/cm at 25°C.

☐ Section 16 ☐
ACRONYMS

ACC	Air Cooled Condenser
APHA	American Public Health Association
AVT	All Volatile Treatment
AVT-O	All Volatile Treatment Oxygenated
AVT-R	All Volatile Treatment Reducing
BOD	Biological Oxygen Demand
CC	Combined Cycle
CFR	Cleaning Force Ratio
COD	Chemical Oxygen Demand
CPT	Coordinated Phosphate Treatment
CT	Caustic Treatment
DCS	Distributed Control system
EDI	Electrode ionization
EDR	Electrodialysis Reversal
EDTA	Ethylenediamenetetraacetic Acid
EPC	Engineering, Procurement, and Construction
EPRI	Electric Power Research Institute
FAC	Flow Accelerated Corrosion
FDA	Food and Drug Administration
GT	Gas Turbine
HP	High Pressure
HRSG	Heat Recovery Steam Generator
HVF	High Velocity Flush
IP	Intermediate Pressure
LP	Low Pressure
MCR	Maximum Continuous Rating
MF	Microfilter, Microfiltration
MIC	Microbiologically Induced Corrosion
MUWTP	Makeup Water Treatment Plant
MWCO	Molecular Weight Cut Off
NAAQS	National Ambient Air Quality Standards

NDE	Non Destructive Examination
NiDI	Nickel Development Institute
NPDES	National Pollutant Discharge Elimination System
NPOC	Non Purgable Organic Carbon
NTU	Nephelometric Turbidity Units
O&M	Operation and Maintenance
OEM	Original Equipment Manufacturer
ONG	Oil and Grease
ORP	Oxidation Reduction Potential
OSHA	Occupational Safety and Health Administration
OT	Oxygenated Treatment
P&ID	Piping and Instrument Diagram
PCV	Pressure Control Valve
PID	Proportional Integral Derivative (Control)
PLC	Programmable Logic Controller
PM	Particulate Matter
RFQ	Request for Quotation
RO	Restriction Orifice
RO	Reverse Osmosis
SAC	Stress Assisted Corrosion
SCC	Stress Corrosion Cracking
SDI	Silt Density Index
SRB	Sulfate Reducing Bacteria
ST	Steam Turbine
TDS	Total Dissolved Solids
TH	Total Hardness
TOC	Total Organic Carbon
TSP	Tri-Sodium Phosphate
TSS	Total Suspended Solids
UF	Ultrafilter, Ultrafiltration
VOC	Volatile Organic Carbon
VPCI	Vapor Phase Corrosion Inhibitor
ZLD	Zero Liquid Discharge

❒ Section 17 ❒

FIGURES AND TABLES

❑ Section 18 ❑
REFERENCES

1. Guidelines for the Operation and Maintenance of HRSGs, HRSG Users Group 2003, TETRA Engineering Group, 110 Hop Meadow Street, Suite 800, Weatogue, CT 06089, USA.

2. L. Bursik, "Seminar Combined Cycle and Heat Recovery Steam Generators was a Big Success," Power Plant Chemistry, March 2002, Volume 4, No. 3, pages 163 – 165.

3. A. Bursik, "Boiler Tube Failures In Industrial Drum – Type Steam Generators," Power Plant Chemistry, March 2002, Volume 4, No. 3, pages 169 – 172.

4. D. Kotwica, "Analysis of Heat Recovery Steam Generator Tube Failures," NACE International, 1440 South Creek Drive, Houston Texas, 77084-4906.

5. J. J. Dillon, P.B. Desch, "Case Histories of Failures in Heat Recovery Steam Generating System," Corrosion 2003, Paper 03-488, NACE International, 1440 South Creek Drive, Houston, Texas, 77084-4906

6. F. C. Anderson, P. S. Jackson,, D. S. Moelling, "HRSG Tube Failures; Prediction, Diagnosis, and Corrective Actions," Corrosion 2003, Paper 03493, NACE International, 1440 South Creek Drive, Houston, Texas, 77084-4906.

7. D. DeWitt-Dick, S. McIntyre, J. Hofilena, "Boiler Failure Mechanisms," International Water Conference 2000, Paper IWC-00-30.

8. K. A. Selby, G.A. Loretitsch, "Water Chemistry Considerations for Operation of HRSG Systems," International Water Conference 1998, Paper IWC-98-55

9. O. Jonas and L. Machemer, "Cycle Chemistry Commissioning Deserves Its Own Strategy," Power Magazine, Vol. 148, No.3, April 2004, pp. 26-30.

10. D. B. DeWitt-Dick, E. S. Beardwood, "Design and Operational Considerations Associated with Chemistry in HRSGs," International Water Conference 2004, Paper IWC-04-30

11. R. Krowech, L. Stanley, "Avoid Damage from HRSG Cycling," Power Magazine, 11 West 19th Street, New York, New York, 10011, March/April 1998 issue, pages 47-50.

12. S. A. Lefton, P. Besuner, G. P. Grimsrud, "The Real Cost of Cycling Power Plants: What you don't know will hurt you," Power Magazine, 11 West 19th Street, New York, New York, 10011, November/December 2002 issue, pages 29-34.

13. Crevice Corrosion Engineering Guide for Stainless Steels, NIDI, www.nidi.org.

14. J. Isaac, L. Gasper, K. Sinha, "From Raw Water Analysis to Steam Purity - Challenges and Solutions," International Water Conference 2003, Paper IWC-03-33.

15. J. O. Robinson, "Chemical Treatment of Gas Turbine Heat Recovery Steam Generators", Corrosion 2001, Paper No. 03492, NACE International, 1440 South Creek Drive, Houston, Texas, 7704-4906.

16. J. Isaac, K. Sinha, "Organic Water Treatment Chemicals versus Ammonia and Hydrazine," International Water Conference 2002, Paper IWC-02-09.

17. Consensus on Operating Practices for Control of Water and Steam Chemistry in Combined Cycles and Cogeneration Power Plants," Water Technology Subcommittee of the ASME Research and Technology Subcommittee on Water and Steam in Thermal Systems, CRTD-Vol. 859988

18. J. C. Bellows, "Steam Purity Recommendations for New Small Steam Turbines", International Water Conference 1995, Paper IWC-95-19.

19. W. H. Stroman, S. Mauldin, D. B. DeWitt-Dick, "HRSG Storage Using Vapor Phase Inhibitors," ASME Research Committee on Power Plant and Environmental Chemistry, 2003 Spring Meeting – West Palm Beach, Florida.

20. M. A. Janick, P. K. Sinha. "Economic and Practical Considerations for Incorporating Condensate Polishers on New Power Plants," 2006, Paper EPRI Eighth International Conference on cycle Chemistry in Fossil and Combined Cycle Plants with Heat Recovery Steam Generators, Calgary, Alberta, Canada.

21. R. Holloway, "The Impact of Water Treatment System Design and Operation on Boiler Feedwater Purity and Boiler Tube Failures," International Water Conference 2000, Paper IWC-00-31.

22. S. J. Geary, J. C. Bellows, "System Monitoring for Chemistry Control," International Water Conference 2000, Paper IWC-00-17.

23. Consensus on Operating Practices for the Sampling and Monitoring of Feedwater and Boiler Water Chemistry in Modern Industrial

Boilers," CRTD-Vol. 81, 2006, American Society of Mechanical Engineers, Three Park Avenue, New York, NY 10016, USA.

24. J. C. Bellows, "Steam Purity Requirements for Combustion Turbine Cooling," Requirements and Methods of Achievement", International Water Conference 2001, Paper IWC-01-19

25. M. A. Janick, "Improved Chemical Treatment Program for Optimum Corrosion Control in a Combined Cycle Power Plant – Prepared Discussion", International Water Conference 2001, Discusser Paper IWC-01-44

26. A Banweg, "Performance Monitoring in the Water Treatment of HRSG's," International Water Conference 1999, Paper IWC-99-14.

27. R. W. Anderson, H. vanBallegooyen, "Steam Turbine By-pass Systems," Combined Cycle Journal, PSI Media Inc., 1917 Hering Avenue, Bronx, NY, 10461 Fall/Fourth Quarter 2003 Edition, pages 3-9.

28. Simon II, David E. The ASME Handbook on Water Technology for Thermal Power Systems, Chapter 21, "Preoperational Protection and Treatment," © 1989, ISBN No. 0-7918-0300-7

29. "Consensus for the Layup of Boilers, Turbines, Turbine Condensers and Auxiliary Equipment," CRTD-Vol. 66, 2002, American Society of Mechanical Engineers, Three Park Avenue, New York, NY 10016, USA.

30. A. D'Ippolito, "Flushing and Chemical Cleaning New Piping Systems," International Water Conference 1989, Paper IWC-89-60.

31. J. C. Bellows, L. Burl, D. M. Smyth, L. McLoughlin, F. Shoemaker, H. Washburn, M. Rizha, "Plant Construction and Commissioning – A Manufacturer's Viewpoint," International Water Conference 2002, Paper IWC-02-10.

32. D. Daniels, "New Pre-commissioning Options for Controlling Corrosion in HRSG's," Power Magazine, 2 Penn Plaza, 25th Floor, New York, New York, 10121-2298, September 2003 issue, pages 49-53.

33. Kuhake, K., R. Ruf, M. Rziha. "28 Chemical Cleanings of Steam Water Cycles During Commissioning", 3rd International VGB/EPRI Conference, Steam Chemistry Interaction of Chemical Species with Water, Steam and Materials During Evaporation, Superheating and Condensation, June 22-25, 1999.

34. Bartholomew R. D. and Hull, E. H., "Preoperational Cleaning Requirements for HRSG Units," PWR 2004-52033, Proceedings of ASME Power, March 30 – April 1, 2004.

35. JJ. M. Sullivan, J. McGraw, "Chemical Cleaning Heat Recovery Steam Generators (HRSG's) Top 11 Lessons Learned," International Water Conference 1998, Paper IWC-98-55.

36. K. E. Hansen, "Pre-Operational Cleaning from a Manufacturer's Viewpoint," International Water Conference 2002. Paper IWC-02-08.

37. D. M. Smyth, M. Rziha, L. McLoughlin, J.C. Bellows, L. Burl, H. Washburn, "Plant Construction and Commissioning – A Manufacturer's Viewpoint of Pre-Operational Cleaning," International Water Conference 2001, Paper IWC-01-10

38. Siegmund, John W. The ASME Handbook on Water Technology for Thermal Power Systems, Chapter 23, "Chemical Cleaning of Utility Equipment" © 1989, ISBN No. 0-7918-0300-7.

39. S. MacDonald, "Preoperational Chemical Cleaning of Heat Recovery Steam Generators," Corrosion 2003, Paper 03582, NACE International, 1440 South Creek Drive, Houston, Texas, 77084-4906.

40. A. Banweg, "Industrial Steam Purity: Requirements, Proper Sampling and Practical Considerations," International Water Conference 2008, Paper IWC-08-28.

41. E. S. Beardwood, 'Pre-Commissioning of Combined Cycle Plants," International Water Conference 2003, Paper IWC-03-13.

42. C. Huth, C. Layman, and K. Sinha, "Challenges in Meeting Condensate, Feedwater, and Steam/Water Quality Limits During Startup and Commissioning of Multi-Pressure Combined Cycle Power Plants – EPC Perspective," International Water Conference 2008, Paper IWC-08-61.

43. GE Energy, "Steam Purity Recommendations for Steam Turbines" GEK 72281f, July 2012.

44. I. Cotton, "Effective Monitoring and Control of Boiler Feedwater Chemistry," International Water Conference 2008, Paper IWC-08-29.

45. American Society of Mechanical Engineers, Performance Test Code PTC-19. 11, "Steam and Water Sampling, Conditioning, and Analysis in the Power Cycle," 2008.

46. Electric Power Research Institute (EPRI), Electricite de France (EDF), and Siemens AG Power Generation, "Flow-Accelerated Corrosion in Power Plants", TR-106611-R1, July 1998.

47. J.O. Robinson, T. Drews, "Resolving Flow-Accelerated Corrosion Problems in the Industrial Steam Plant," CORROSION/99, paper 99346, NACE International, Houston, TX (1999).

48. ASM International Handbook, "Corrosion: Fundamentals, Testing, and Protection", Volume 13A, 4th Edition, October 2008, page 210.

49. Macdonald, Digby D. and Gustavo A. Cragnolino. The ASME Handbook on Water Technology for Thermal Power Systems, Chapter 9, "Corrosion of Steam Cycle Materials" © 1989, ISBN No. 0-7918-0300-7.

❑ Appendix A ❑
DAMAGE/CORROSION MECHANISMS

Caustic Gouging – The dissolution of carbon steel by localized high concentrations of sodium hydroxide. This high-pH condition solubilizes the steel's protective magnetite layer and ultimately forms sodium ferroite ($NaFeO_2$) and sodium ferroate (Na_2FeO_2). In-situ corrosion products remain in place and smoothed surfaces typically are present below the deposits in the attacked area.

Corrosion – Degradation of a material and its properties resulting from chemical or electrochemical reactions within the environment.

Corrosion Fatigue – Life reduction due to the occurrence of cyclic applied stress (fatigue) in a corrosive environment. Component lifetime under corrosion fatigue is less than if fatigue life and corrosion life had both been assessed separately because of continuous breaking of the protective oxide film by fatigue action.

Creep – The time-dependent, thermally-assisted deformation or strain of components under load (stress). Creep damage in a given material is a function of a combination of time, temperature, and stress, and results from a combination of high stress and high temperature relative to a material's creep limits. Damage occurs in three stages. Visible damage is induced during the tertiary stage. Isolated creep microvoids coalesce along grain or dendrite boundaries, resulting in fissure formation prior to stress rupture.

Crevice Corrosion – Localized corrosion resulting from the formation of a concentration cell in a crevice formed between two surfaces that shield it from the full oxygen content of the environment.

Fatigue – Repeated applied stress cycles that result in reduction of strength, crack initiation, and crack propagation. Most of a component's fatigue life occurs prior to crack initiation. Crack morphology varies based on type of fatigue loading (i.e. high-cycle, which typically is induced by vibration, or low-cycle, which is typically induced thermally) and environment (for example, corrosion fatigue cracks often are oxide-filled and produce secondary parallel cracks).

Flow-Accelerated Corrosion (FAC) - FAC is a phenomenon that affects the normally protective oxide layer formed on carbon or low-alloy steel. With FAC, the oxide layer dissolves into the flowing stream of water or a water-steam mixture. As the oxide layer becomes thinner and less protective, the corrosion rate increases. Eventually a steady state is reached where the oxide dissolution rate equals the oxidation rate, and no oxide exits on the metal to slow corrosion. It is important to note that in the FAC process, the protective oxide film is not mechanically removed. Rather, the oxide is dissolved. Thus, FAC may be defined as corrosion enhanced by mass transfer between a dissolving oxide film and a flowing fluid that is unsaturated in the dissolving species[46,47]. This mechanism is also known as flow-assisted corrosion or flow induced corrosion.

Fretting; Fretting Corrosion – Fretting refers to metal deterioration caused by repetitive slip (on the order of microns) at the interface between two contact surfaces. Small metallic particles spall during metal-to-metal contact and continue to induce damage. This mechanism is often called fretting corrosion because of the red oxide produced by oxidation and fretting when one or both alloys are ferrous-based.

Galvanic Corrosion – Corrosion caused when a metal or alloy is electrically coupled to another metal or conducting nonmetal in the same electrolyte. When dissimilar metals are electrically coupled, the current flow between the metals causes increased corrosion of the less corrosion resistant metal, and decreased corrosion of the more corrosion resistant metal[48].

Gas-Side Decarburization – A high-temperature gas-metal reaction whereby carbon (typically in the form of pearlite or iron carbides) is selectively removed from the surface of a metal alloy. High concentrations of carbon dioxide can combine with iron carbide in the steel to form iron and carbon monoxide at high temperatures. Reduction in the carbon content typically reduces surface strength.

Gas-Side Dew Point Corrosion – Corrosion caused by the formation of acids, typically sulfuric acid, when the combustion flue temperatures drop below the condensation point for the acid present as a vapor in the flue stream. This condition can cause pitting corrosion during shutdowns of equipment that typically sees elevated temperatures in service. Such corrosion most often occurs during the shutdown of components which operate above dew point temperature.

General Corrosion – Non-localized corrosion that occurs over a large percentage of the metal surface, resulting in uniform metal loss. Seawater flowing over carbon steel, for example, can generate general corrosion.

High-Cycle Fatigue (Vibration) – Fatigue associated with low-amplitude, high-frequency compression - relaxation cycles from vibrational loading.

High Temperature Oxidation - Also known as thermal oxidation, this is a reaction where a metal or alloy is converted to an oxide due to exposure to elevated metal temperatures in an environment that allows the metal to combine with oxygen at excessive rates.

Hydrogen Damage – A general term for the embrittlement, cracking, blistering, and hydride formation that can occur when hydrogen is present in some metals.

Liquid Droplet Erosion - A wastage mechanism caused by impingement of droplets of a liquid entrained in a vapor or gas stream at high relative velocity to a solid surface.

Microbiologically Influenced Corrosion (MIC) – MIC is an electrochemical process in which the source is biological. MIC is often caused by sulfate-reducing bacteria, but other biological initiators also exist. Chemistry changes can be brought about by the depletion in oxygen and the production of enzymes and waste metabolites. Direct metal dissolution can also be incurred due to oxidation-reduction reactions of the microorganism's metabolism. MIC can occur at temperatures up to 235 °F, and often occurs in stagnant flow conditions.

Pitting Corrosion – Corrosion of metal surfaces confined to small points or areas that takes the form of cavities in the surface. Pits result from localized galvanic cells which often are associated with local discontinuities including inclusions, local deposits, or local composition deviations.

Phosphate Gouging – The dissolution of carbon steel by localized high concentrations of mono- or di-sodium phosphate to form maricite ($NaFePO_4$) and other sodium-iron-phosphate compounds. Crusty alternating black-and-white layered deposits may remain with knife-edge like surfaces in the gouged area. This form of corrosion also requires a phosphate concentrating mechanism, such as steam blanketing or boiling within porous deposits, and the development of a thermal barrier between the tube metal and the boiler water.

Stress-Assisted Corrosion (SAC) – Corrosion caused by a combination of high local stress, either applied or residual, and a corrosive environment. Highly localized stress can cause breaks in the protective oxide film, so that the presence of a corrosive environment initiates attack where the film is broken. Often the attack appears as cracks composed of a series of interconnected pits. (SAC is a process in which either an applied stress or residual stress is present that causes a break in the protective oxide that does not self-repair at a rate fast enough to prevent the corrosive environment from accelerating the corrosion).

Stress Corrosion Cracking (SCC) – A cracking process that requires the simultaneous action of a corrodent and sustained stress acting on a susceptible material. Cracking damage can appear as intergranular or transgranular, multi-branched cracks, depending on the alloy and the environment. Cracks are typically non-oxide-filled and branched along their lengths.

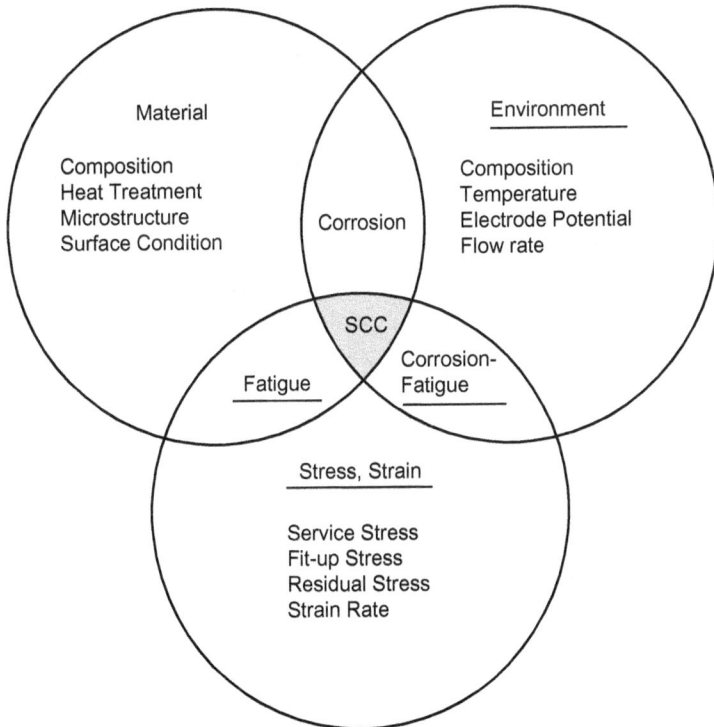

Figure 4:   Three Groups of Major Factors Determining Susceptibility to Stress Corrosion Cracking[49]

Three groups of major factors determining susceptibility to stress corrosion cracking (SCC). Specific values are required within each group for SCC to occur. In contrast, corrosion, fatigue and corrosion-fatigue require specific values in only two groups of factors.

Thermal Fatigue – Fatigue associated with cyclic temperature changes that cause the metal to expand and contract. The mechanism is considered to be high-amplitude, low-frequency low-cycle fatigue.

Under-deposit Corrosion – Corrosion damage mechanisms where apparent surface deposits are present and damage occurs under the deposits. The deposit sources may have come from upstream or may have been generated by the corrosion mechanisms themselves. These mechanisms can include mechanisms such as caustic gouging, hydrogen damage, and phosphate corrosion.